I0466557

HUMANS

The Last Endangered Species

HUMANS
The Last Endangered Species

Edward J. Ryder

MBD
PUBLICATIONS
A Division of MONTEREYBAYDESIGN

HUMANS: THE LAST ENDANGERED SPECIES

MBD Publications
A Division of Monterey Bay Design
Salinas, California

Cover Art and Interior Design by Deborah Ryder

ISBN-13: 9798333517487

Printed in the United States of America.

1 3 5 7 9 10 8 6 4 2

Table of Contents

Introduction

When some early human beings began to think about the time periods past, present, and future, a few of them may have wondered whether it was possible to make changes: to alter happenings that had already taken place, those that were occurring, and those that had not yet occurred. Obviously, these are different phenomena groups, whose nature must be assessed in different ways. Within the groups, there may be substantial variation in the nature of the happenings, so that each may have to be examined and/or treated differently, if at all.

Past occurrences may be acceptable and need no modification. They may not be acceptable, and the question may be asked: Can the results be remedied if necessary? If so, steps may be taken. If not, one may live in despair or adapt to the changed circumstances.

Occurrences in progress may be interrupted and redirected. If that is not possible, one may live with the consequences, or perhaps run away. Living with the consequences may enable one to continue to attempt to make more desirable changes.

If it seems that undesirable happenings will occur in the future, actions may be taken that might prevent the worst results. The wording of these paragraphs does not describe actual events or prospects, which await the reader in the essays of this book, framed in the overall concept: threats to human survival.

Many stories of the future have been created in books, movies, radio, and television. Overwhelmingly, they usually have concluded with some version of a happy ending, but that ending might have been in the ashes of a catastrophe.

This volume requires an introduction that includes the author. A perusal of the table of contents may suggest that I consider myself an expert on everything. Not so. My professional field includes genetics and plant breeding. These include familiarity with several vegetable species, especially lettuce, which was the crop I researched while employed by the U.S. Department of Agriculture. That does not mean that I am relatively ignorant of everything else. Rather, eighteen years of formal education, plus many years of paying attention to various life experiences: reading, watching, and listening to various sources of information, have contributed to a broad knowledge of the subjects discussed in this book. The reader, of course, may be the ultimate judge of my competence in making use of those experiences.

The direct threats of disaster are discussed in the first three essays. These are the human genome, war, and climate change. Each can be a direct threat to attack our existence. The remaining essays discuss the various human institutions and systems that can cause the materialization of the threats: how that has happened in the past, is happening in the present, and may well happen again in the future, with dire consequences.

It is my hope that the specific combinations of information and ideas presented are relevant and unique to readers of this book and hopefully will be transmitted to others and eventually to various world leaders who might be persuaded to act and save our civilization from itself. This may sound like a tall order, and it is. But I believe the time is ripe to consider the possibilities.

In fact, the time has been ripe for at least three fourths of a century. Earlier than that, we were aware that the population was increasing, but not well enough aware to be seriously concerned. In 1945 we were confronted with the frightening reality of the most powerful weapons ever: nuclear bombs. About forty years ago our attention was drawn to the reality of climate change and global warming. Various forms of pollution, of the air, the water, and other forms, had been facts of life for centuries. Each of these phenomena posed varying degrees of danger threats. Together they warn us: "We are the components of Armageddon. Beware!" I hope that the readers will take this warning seriously. I also hope they will demand that the peoples of the many countries of the world also take it seriously and demand action to prevent the worst from happening.

There is one important factor that should be kept in mind throughout the pages of this book. The word "we" does not usually mean all of us, but rather some proportion of the population of the world, which may be large or small depending upon many factors, especially the genetic makeup of the persons being described. Therefore, it should be kept in mind that for the traits that will be described in the essays, each person will have a different set of genes from all others, excepting identical twins. That characteristic will exert great influence on the type of behavior, level of intelligence, and other relevant human traits, and will strongly influence the behavior of people. These statements will be made in greater detail appropriate to the specific subject.

1

The Basic Threat: The Human Genome

Our little world is sore all over, constantly battered with blows from every imaginable source. Storms, earthquakes, volcanoes, wildfires, stifling heat, freezing cold, and other natural events come with the territory. One might have hoped that we humans, filled with compassion for our suffering planet, would do a better job of ministering to its woes, treating it kindly, petting it, and saying "There, there!" occasionally. Instead, we have spent many of our three hundred millennia on its surface doing our own bashing, thus adding injury to injury. We have endured thousands of years of almost constant warfare, killing each other senselessly, and spilling our blood on its surface. Pumping debris into the air. Tearing apart great swaths of its beautiful surface. Demolishing its life-giving forests. We are trying very hard to destroy this planet, along with ourselves. Why are we so stupid?

We ought to consider two questions. What is the cause of our destructive behavior? What, if anything, can we do to change course and set about constructing a better world? I will examine several human activities that should explain the relationship between our genes and those activities.

Regarding the first question, the trouble is primarily in our genes. This shouldn't be a surprise. Our genes define us, driving our thoughts and actions, many of which, unfortunately, cause harm. Over the years, we have found new and better ways to cause ever-greater damage. Eventually, urged on by our genes, we may unleash a world disaster that will be unredeemable. Sound preposterous? Read on.

First, let's consider some basic information about genetics and evolution, the driving forces of human growth, development, and behavior. The many other species on earth are also subject to these forces, which have led to the tremendous variety of creatures that exist on the planet. The collections of genes in our bodies initiate the development of the traits that define us, individually and as a species, while evolution has been the driving force behind change over time, selecting among the genes that eventually created several pre-human species and most recently, our species, *Homo sapiens*.

Neither genetics nor evolution seeks perfection because these forces are mindless. They have no idea what they are doing. Genes change, through a process called mutation, in a random, non-purposeful manner, producing forms with varying characteristics. Evolutionary selection among the various forms is usually based on their relative fitness to survive in an existing environment. Evolution has no goals. We humans should not think of ourselves as the end products of a long-term drive towards biological perfection, at the top of the tree of life, supreme over all other forms on the earth. Our species is merely one branch out of thousands on the same tree.

However, we are the most intelligent, adaptable, and innovative species on earth, and therefore the most dominant. Evolution has bequeathed us the genes in our cells that control our intelligence, our actions, our personalities, and our needs. We should pay more attention to the consequences of our existence and our abilities to do things well or badly. We are the beneficiaries of our genes' actions, but unfortunately also their victims and their instruments of harm.

How does this happen? First, there are several important things to know about genes to understand why they are important to each of us, individually and as members of the human society.

So, let's look at some additional basic facts. One is that genes reside in our body's cells, on chromosomes, which live in a part of the cell called the nucleus. The chromosomes, and therefore the genes, occur in pairs. The members of a gene pair are called alleles. One allele is on one chromosome of the pair and its mate on the other chromosome at the matching location. Together they produce an effect on the organism, which is called a trait. Eye color is a trait, and height, and many others that make us what we are. A qualitative trait results from the action of one gene, a single pair of alleles. Quantitative traits are those controlled by several to many genes acting together. Quantitative traits are more complex and therefore more difficult to analyze genetically than qualitative ones. Their effects are slightly or substantially modified by environmental influences. Such traits exhibit a wide range of expression. They are our most meaningful and influential characteristics.

Another fact is that human genes get around. They

travel in and among populations in two ways: vertically, from generation to generation through mating, and horizontally, through migration plus mating among different, sometimes widely dispersed populations. Consider, for example, that Genghis Khan and his conquering Mongol hordes left their genes all over Western Asia and Eastern Europe.

Each person has about twenty thousand coding genes, which create proteins that lead to various traits. The genes occasionally mutate, modifying their functions and the traits affected. These phenomena create enormous variation among humans, over time and across borders around the world. As a result, no two people have the same genetic makeup and the same trait expression, except identical twins. Furthermore, it is usually difficult to predict when, where, and in whom specific genetic combinations will show up. Let's explore the consequences of these actions.

Human genes perform four types of jobs: a) control physical attributes; b) control reaction to disease organisms and disorders; c) control intelligence; and d) control behavioral traits. Behavioral traits and intelligence, especially, define each person's inclination and ability to find a place in the world and to impact on local, national, and world affairs.

We study the inheritance of traits in plants and many animals by using standard crossing techniques between selected sets of parents followed by analyses of their progenies. We can then estimate the genetic and environmental contributions to traits. In human genetics, directed mating is not usually an acceptable practice, so we use other techniques, such as twin studies and analyses of family histories. These techniques have been useful in intelligence studies and in some behavioral studies, such as schizophrenia and bipolar disorder. Most of the behavioral traits mentioned in this chapter are likely quantitative in nature and therefore would require difficult and complex analytical techniques to thoroughly explore the nature of their inheritance. The research done so far would support the statement that there is a genetic basis for behavioral characteristics, with a strong environmental influence for some and a weaker one for others.

Of the four types of genetic traits, physical genes control attributes of the body, such as strength, mobility, functioning of body parts, and various aspects of appearance. Those alleles that are beneficial to the body are "good". Alleles that are deleterious are "bad". The alleles of some traits, such as for ordinary height and certain eye color differences, are neither good nor bad because the traits themselves are not good nor bad, just different, such as eye color.

Some health genes contribute to the body's reaction to diseases caused by viruses and microorganisms. A person may be resistant or susceptible to those diseases. Others contribute to various disorders of the body, such as cancers, multiple sclerosis, diabetes, and heart disease. For a trait controlled by quantitative genes, the relative numbers of good and bad alleles contribute to the incidence and severity of the malady.

Intelligence genes control our ability to learn, solve problems, create new things, and explore the unknown, enabling us to continually expand our knowledge and accomplishments. Various environmental forces (learning opportunities, peer pressure, illness, etc.) also contribute to intelligence level. Research has shown that the genetic contribution to intelligence is almost identical in each member of identical, or monozygotic, twin pairs and less so in fraternal, or dizygotic, twins. This is true even when identical twins are raised apart, in different environments.

The genetic contribution to intelligence is substantial, stable, and real.

Behavior genes affect our needs and actions. Our behavior may affect others and influence small or large segments of a population, and the effects can be good or bad. I have named the following fifteen traits, a few among many that define a person's behavior, with a range from the detrimental extreme to the beneficial one. The names of the traits are selected by me and may not match the names used by current professional geneticists. Nevertheless, the descriptive words should be reasonably recognizable.

Behavioral traits and intelligence, together, define each person's inclination and ability to find a place in the world and to have an impact, beneficial or harmful, on local, national, or world affairs.

TRAIT CATEGORY	DETRIMENTAL	BENEFICIAL
Craving	Greediness	Generosity
Attitude	Arrogance	Modesty
Attitude	Aggressiveness	Friendliness
Acceptance	Intolerance	Tolerance
Belief	Credulity	Skepticism
Reasoning	Irrationality	Rationality
Action	Violence	Gentleness
Attitude	Uncaring	Caring
Communication	Tells Lies	Truthful
Action	Cruelty	Humaneness
Craving	Selfishness	Altruism
Wariness	Suspicion	Trust
Attitude	Jealousy	Admiration
Ambition	Extreme	Moderate
Ethics	Dishonesty	Honesty

Let's examine behavioral traits more closely, including their effects on human motives and actions, and consider the possible genetic bases for their expression. If we assume that these traits are quantitatively inherited, the expression will vary in the population from one extreme through intermediate categories to the other opposite extreme. For example, a person may be very greedy, moderately greedy, or slightly greedy, ranging to slightly, moderately, or highly generous. Similar variation in range occurs for the other traits named above. Each of the traits can vary independently, yielding many combinations for each trait and among all the traits, at many possible levels. Except for identical twins, no two people will be the same, but they may be quite similar or markedly different. In addition, certain trait combinations may supplement each other: Person A, who is ambitious, aggressive, greedy, and intelligent is likely to have a strong desire for money and power. The desire for power is a varying need in the human species.

Person B, who is ambitious, but also altruistic, may desire a more benevolent type of leadership in science, education, politics as a means of service, or some other occupation less dependent on money and power. Benevolent leaders who may acquire power, but without the consuming hunger for it, view this power as a great responsibility and an opportunity to enhance the lives of large numbers of people.

For those like Person A, the desire for power is a basic need, although people who are less ambitious may have less need for power and may shy away from a high level of leadership. For those that actively desire power, there is a wide range of need, from a modest ambition to be president of a local club or foreman of a shop to a consuming hunger to acquire great riches, run a corporation big enough to swallow competitors whole, become the absolute ruler of a country, or conquer the world.

The direction a person takes in the search for power is dependent first upon the highly variable distribution of alleles that fuel the need. Such persons must also have sufficient intelligence, exist in a succession of environments that support that need, and be able to take advantage of opportunities that arise. For people with a great craving for power, the lesser opportunities are only stepping-stones to positions of enormous authority, which offer chances to do things that affect large numbers of people over the entire world. Leaders of countries and global corporations may affect the lives of thousands, millions, or even billions of people.

Benevolent leaders, harboring a supportive mix of mostly good alleles, who may acquire power without the consuming hunger, view this power as a great responsibility and an opportunity to enhance the lives of large numbers of people. For leaders who worship power, the opportunity usually means personal gratification, often with disregard for the effects on others. Extreme expression of aggressiveness, avarice, cruelty, violence, destructiveness, and irrationality may create a person capable of unleashing disaster. Combined with sufficient intelligence, these traits may produce a world-class monster. Three people of that description emerged in the twentieth century alone: Adolf Hitler, Josef Stalin, and Mao Zedong. What havoc could have been visited upon the world if the super conquerors of the past, such as Genghis Khan, Alexander the Great, or Napoleon Bonaparte had the technological tools that are available now? And what will be available to future leaders of this ilk?

We humans have existed for about three hundred thousand years. During that time, we have populated the earth and covered it with the things we have created. No other species has done that. All our collective actions and accomplishments, good or bad, have had leaders, except possibly in the earliest human groups. Some leaders have been decent; some have been bad; and some, but far too many, have terrorized large sections of the world.

Many people who are not leaders may have some of the alleles for cruelty or indifference or other traits but lack those for ambition or aggressiveness. Their actions will affect a relatively few people—family, friends, or colleagues. They may also become followers of those with the greater hunger.

Power! Power hunger, fueled by the wrong behavior alleles, guides some leaders and authorities at all levels: the prison guard who bullies prisoners; the business owner who mistreats employees; political leaders who can be purchased; business and financial leaders who accumulate billions by essentially stealing from clients, customers, employees, and/or the general public; rulers, whether presidents or kings, who make war on their citizens and on their neighbors; tyrants who conquer and destroy many, many people and the trappings of their civilizations. All of them work under the thrall of the alleles that fuel the need for more power, more money, more subjects, or all of these. Some of the bad alleles, such as those for greed and indifference, also incite the fouling and plundering of our planet's resources.

Bad human alleles have probably been around as long as the human species, and they are not going away. They cannot be excised, not even by the horrible eugenic methods that have been perpetrated on great numbers of powerless people through segregation, forced sterilization, murder, and starvation. They have and will pop up continually in succeeding generations, sometimes where least expected. Alexander the Great was the son of the king of Macedon, and

Genghis Khan was the son of a tribal chieftain, but Hitler was the son of a minor public official in Austria, Mao was a farmer's son, and Stalin's father was a cobbler from Georgia.

To repeat the earlier words of caution: The discussion up to this point has been based on limited but increasing evidence plus an educated and realistic *assumption* that there is a measurable, substantial genetic basis for all behavioral traits. Let us further examine the evidence that genetics *is* that basis. All that we know of the behavior of Hitler, Stalin, and the other leaders of history, or of any of the current political, corporate, educational, scientific, religious, or other leaders stems from reading about them and/or observing them, acquiring information from which we may try to explain their actions.

To enable us to proceed past assumptions and speculation, we must test the following null hypothesis: Genes do not influence the behavior of human beings. We continue to conduct studies on specific human behavioral traits to confirm that heredity is responsible for them, as it is for intelligence. Human general intelligence is also quantitative and must be measured indirectly. We understand intelligence at a level that allows us to reach tenable conclusions. The primary tool for our understanding of the inheritance of intelligence has been the study of twins, comparing IQ test results of identical twins raised together or separately, fraternal twins, and non-twinned siblings. Recent molecular research offers additional tools that confirm some of the twin study results. These studies have shown conclusively that heredity plays a major role in intelligence. Similar methods apply for studying traits of behavior, and the new molecular techniques will enable us to directly identify certain genes. We must continue to avail ourselves of these techniques. That null hypothesis, therefore, is probably wrong.

The overall distribution of the alleles of behavior genes in the human population may be partly random, and partly due to familial trends. The number of genes controlling each trait may range from one to many. The total number of relevant genes, encompassing several traits, is very large. Consequently, the number of behavior genotypes must be very large and include "very good" people, "very bad" people, and every gradation in between. Their distribution in the human population probably approximates a normal curve, also known as a bell-shaped curve. (See illustration at the end of this essay) Each genotype's effect is moderated by various environmental influences and some sequence of opportune situations. Roughly the same proportion of people with the various distributions of good and bad alleles will probably exist in succeeding generations. Selective forces that change the distributions will operate over very long periods.

The study of behavior genetics began more recently than that of intelligence. Much of the research has explored certain personality disorders, such as schizophrenia, bipolar disorder, major depression, and autism. A few more recent studies have found probable genetic bases for several traits: jealousy, superstition, violence, ruthlessness, and aggression, and indicate they are probably quantitative effects. The methodologies employed to study these phenomena included twin studies using various survey procedures plus molecular studies.

As geneticists develop a solid foundation of information on the inheritance of behavioral traits, we then may consider what, if *anything*, can or should be done about the various expressions of human conduct. Many questions may be asked. The answers will not be simple, may not be legally and morally acceptable, and may not be possible to carry out. With this daunting statement, let's consider some possible questions.

First, what does it mean to understand the inheritance of human behavior? The methods in use now are designed to understand behavior on an overall group basis; they do not disclose the actual gene distribution in individuals. The information derived from a set of survey questions is that a certain proportion of the variation among the group members is due to genetic causes and the rest is from various environmental effects. This tells us nothing about the distribution of behavior alleles in each person.

What do we know at present of human genetic information? On one hand we have identified genes in terms of their effects, and the characters have been named: in insects, plants, and non-human animals, as well as humans. This process is dependent on Mendelian genetics. More modern genetics research is providing maps of gene locations on DNA strands. This information is not useful alone for those who are doing practical research such as plant and animal breeding or medical treatments. Matching the Mendelian and map information should be a useful tool in various types of research.

For example, can genetic information about individual persons be ascertained? Or, in other words, can we augment the map of the human genome, showing the location and individual identity of all genes affecting traits such as height, reaction to diseases and disorders, eye color, hair curliness, and foot size, and those that control the level of intelligence and a full range of behavioral traits? Can we learn to predict whether a person will be generous or greedy, indifferent or caring, violent or gentle, arrogant or modest, and so on? The answer is: Quite likely. When will these things happen? Research is in progress now. I must point out that nothing is known of the actual behavior genes of the villains named earlier, only that they were bad persons, and it is likely that they had an array of bad behavior alleles.

It is one thing to fully understand the specific genetic makeup of people. What would be the consequences of using this knowledge to change a person's genes? Let's consider a hypothetical situation. We may discover that 10% of a test population are very greedy, 10% are very generous, and all others are somewhere in between. Or that 5% tell a lot of lies, and 5% nearly always tell the truth, while again all others are somewhere in between. We may also discover that 0.5% combine aggressiveness, ambition, indifference, and cruelty. And so on for all the traits I have named, individually and in combinations, as well as any others that might be considered.

Now let us suppose that this information specifies the exact genetic makeup of each person and can be made available to the parents of a newborn child, and perhaps also to schoolteachers, prospective employers, various government institutions, etc. How can this information be used? We are immediately confronted with a word: eugenics. Its definition sounds benign. It is a social philosophy that advocates the improvement of the genetic quality of the human population by promoting higher rates of reproduction for people with desired traits and/or lower rates for those with undesired traits. This statement deserves further exploration, which is taken up in other essay discussions.

It is obvious that much more work needs to be done, particularly to discover broad quantitative genetic bases for behavior traits to establish that the variation among humans is normal, with a range of differentiating effects, as with intelligence and many physical traits. Studies of single gene effects often disclose extreme manifestations of a trait, such as giant and dwarf states as compared to the normal range of height. Once we have developed a solid foundation of information on the inheritance of behavioral traits, including those named above, we will then be able to consider what, if anything, can or should be done about the quest for power and its abuse. It is important to understand that studies of this sort are and will be on an anonymous group basis, so that individual identities are not disclosed.

I find it hard to be optimistic about the future. The sword over our heads is the prospect that the use and abuse of power and money by actual and aspiring political and corporate leaders will continue to become progressively worse, together with an inexorable movement in the direction of a global disaster from which there may be no return. What sort of disaster? We can consider the following: Global warming to the point where most of the ice melts and many of the world's shorelines are inundated. Or, an ever-widening wealth gap, possibly leading to worldwide economic collapse. Or, nuclear warfare, resulting in immediate mass murder in places where the devices explode, and delayed mass murder where radioactive fallout descends from the sky. It is not unlikely that we might absorb all the above in a perfect planetary storm. After all, we have been waging war for thousands of years, and getting better at it. We have been toying with inducing economic collapse at least since the invention of the stock market in the Sixteenth Century. And the seeds of global warming were probably planted during the industrial revolution.

I have explored human genetics and evolution as controlling forces of the nature of leadership, emphasizing certain quantitatively inherited traits, especially intelligence and various behavioral characteristics. I have also considered several societal institutions that may become threats to our very survival. Four of these institutions have been created as or have become instruments of threat: war, global warming, pollution, and over-population. Others were developed to provide us varying indicators of civilization, such as science, technology, business, agriculture, education, belief systems, communication, government, politics, and inequality.

In the following essays, I will explore each of the above phenomena in additional detail to illustrate the dangers that we will continue to face.

NORMAL OR BELL-SHAPED CURVE

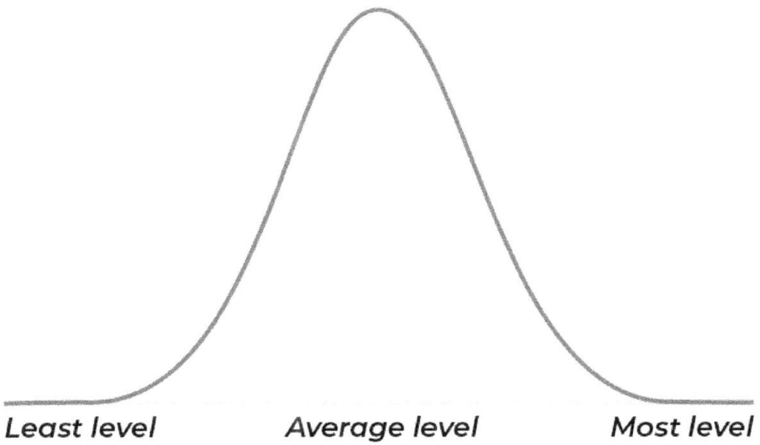

Heights along the curve indicate the number of persons at that location.

2

War and Its Weapons

Somewhere in Africa several million years ago, the first hominid to absorb a body blow from a rock, stick, or bone did not likely realize the historical notability of the occasion. He knew only that it hurt, and that he was either angry or afraid or both. Was it an act of war? If so, was it a rebellious blow, struck by a member of his own group, or an attack by a rival group in a dispute over food or territory? I am assuming that the antagonists were male, presumably the more aggressive members of the groups. While reading this discussion, the reader should keep in mind that the differences and similarities are based on genetic differences and similarities among all the people participating in the conflicts.

Certainly, none of the participants in the melee that probably followed that first blow could begin to imagine the scope of warfare in the distant future that would threaten human existence itself. As time passed, additional weapons and equipment, such as knives, spears, darts, bows and arrows, horses, and chariots came into use. About the year 5000 BCE the Bronze Age began, featuring the development of metal weapons such as daggers and swords. The rate of weapon development accelerated from tens, hundreds, or thousands of years between advances to a comparative equivalent of the speed of light. After the invention of gunpowder and slow firing weapons such as muskets, the inventing process produced pistols, rifles, machine guns, cannons, bombs, rockets, and mobile weapons for use on land, sea, and air.

There probably was a time when conflicts started accidentally, perhaps when two groups, in search of better food and water sources happened to arrive at the same promising location at the same time. Both groups may have been tired, hungry, and thirsty, and unwilling to allow the other group ownership of the location. A "war" may have ensued, with whatever weapons were available: clubs, stones, bones or just fists and feet. Eventually, one group probably would have given up and run away or been destroyed, unaware that they were making history.

A slight difference in timing might have created a situation where the conflict would have been purposeful rather than accidental. If one group arrived at a location, observed the favorable signs: water, food, and shelter, they might have decided to stay. Then, if a second group arrived later, hungry and tired, their decision might also be to stay. They might then have attacked the first group to take over the location. This would have been a deliberate conflict. There we have the establishment of two different classes of warfare, based on different but similar situations, and a forecast of the many kinds of conflict to come, albeit with more dangerous weapons, and possibly for less vital reasons.

Wars probably remained this simple for a very long

time as the speciation of hominids continued, until the appearance of *Homo sapiens* about three hundred thousand years ago in Africa. Some of the early sapiens groups migrated north and then into Europe, the Middle East, and southern and eastern Asia. About ten millennia ago, humans invented agriculture, leading to the almost complete disappearance of the hunter-gatherer culture, and the growth of villages, towns, and cities.

These developments brought on an increasing complexity to warfare, weapons, and the reasons for engaging in conflict. The appearance of cities was a signal development in the concepts of settlement, possession, expanded kinship, and permanence of place, that didn't exist for the wandering ancestors of the new groups of dwellers. The concept of "us against them" may have arisen in their minds, leading to the first conflicts, likely of a minor nature, that may have arisen between people from two neighboring settled populations.

The minor conflicts would soon become major ones. Early on, they apparently took place in Mesopotamia, a section of Asia east of the Mediterranean Sea. The first historically documented war was in the year 2700 BCE, between Sumer and Elam, which was won by Sumer. Thus, after the first 300 millennia, war became an "official" occupation of humanity, with increasing levels of complexity, danger, and destruction. The lands surrounding the Mediterranean Sea, as well as the sea itself, hosted wars of conquest involving Grecian city-states against each other or against Persia and other Mediterranean city-states.

In the 8th Century BCE, Rome engaged in the first of dozens of wars as a republic and later as the Roman Empire. Rome fought its mostly successful wars against Carthage, Gaul, Britain, and Parthia. However, Gothic invaders finally ended the glory days of the Western Roman

Empire, prior to the Middle Ages. Series of wars were also fought in other parts of the world: China, the Americas, and Africa. Wars have continued to be a part of human culture to the present day. They have been fought all over the world, from minor skirmishes to world conflicts. There have been so many that they can be classified in multiple ways: by number of participants and casualties, by purpose, and possibly even in the Linnean fashion, like plants and animals.

War is a means to do harm to others. It is a means to punish others. It is a means to take things from others. It is a means to acquire power from others. It is a means to own others: their bodies, their lands, their belongings, their rights. It is a means of doing business. It is a means to settle arguments. War is also a means of response to each of the above goals.

War has consequences. Most devastating are death, physical injury, and emotional damage to those fighting. In addition, there is destruction of buildings, sometimes including entire cities, and the inflicting of terror, death, and injury even on those not participating in the conflict. War has lasting effects beyond the termination of the strife: the struggle to recover from the loss of loved ones, the fear, the struggle to rebuild, and other traumas. Some of the results have become useful and good for society, such as inventions and discoveries that might never have happened otherwise.

War is terrible. Reasons for going to war arise and are claimed to be legitimate by the perpetrators, but by and large, those reasons are usually better described as excuses. Perpetrators is a four-syllable substitute for men, overwhelmingly the responsible parties for each new war. In discussing many kinds of behavior in humans, this is a key definition, as the reader will discover.

Turning to the classification of wars, based on the

ostensible reasons for making war, let's examine some of the principal types of war: neighborly attacks and responses, attempts at a series of conquests, religious conflicts, acquisition of wealth, colonialism, war as a business, civil wars or revolutions, and world war.

Neighborly wars have been common, for various reasons: proximity, plus a desire for more land, a better castle, more slaves, access to waterways like a river or a sea or an ocean, or power and glory. Several wars have been fought to satisfy a "need" for large-scale conquest, successively attacking multiple city-states or countries. These have been perpetrated by historically famous conquerors, such as Alexander the Great, Julius Caesar ("I came, I saw, I conquered"), Genghis Khan, Tamerlane, Attila the Hun, Napoleon Bonaparte, Adolf Hitler, and Joseph Stalin. Conquerors like these men often become famous, because they seem bigger than life while basking in the glory of the military. This assessment seems ridiculous, and certainly evil, when we consider that their mission in life is the death of others and massive destruction.

Religious differences have been principal causes of war for a long time, as well as an underlying basis for some wars that were fought primarily for other reasons. The best-known conflicts for which religion was primary include the Crusades, pitting Christians from European countries against Muslims from states around the Eastern Mediterranean Sea. In France, wars between the Catholic monarchy and Huguenot (Protestant) princes were fought in the latter half of the 16th Century. The Thirty Years War (1618–1648) was one of the most devastating series of conflicts in history, combining a religious factor involving Catholicism, Lutheranism, and Calvinism with territorial designs encompassing most of the European states.

Religion has played its part even in the 20th and 21st Centuries. Conflicts between India and Pakistan trace back to the withdrawal of British occupiers from the Indian subcontinent after World War II. The Muslim leadership insisted on partition from India, and the transition itself was bloody. Conflict has continued intermittently with repeated bloodshed and threats to become nuclear. In the Middle East, four wars were fought between the new Jewish state of Israel and its Muslim neighbors: 1948 Arab v Israeli War, 1956 Suez Crisis, 1967 Six-Day War, and 2023 Hamas v Israel. The strife continues. In addition, simmering rivalry between the Shia and Sunni Muslim states intermittently breaks out into armed conflict. Finally, violent sects such as Al Qaeda and ISIS essentially declared war on all infidels, in other words, everyone else on earth.

Two world wars in the 20th Century involved multiple countries from all populated continents. World War I was triggered in 1914 by the assassination of Archduke Franz Ferdinand of Austria-Hungary by Gavrilo Princip, a Serb who was part of a group that desired to consolidate the Balkans as the country of Yugoslavia. War ensued when Austria attacked Serbia, and Germany attacked Serbia and Russia. France and Great Britain declared war on Austria and Germany. The United States, after losing some ships to German U-boats, joined the conflict in 1917. Many countries from all over the world contributed to the war, which ended by armistice in 1918. It was dubbed as the war to end all wars, a goal supposedly guaranteed by the League of Nations, newly formed in 1920.

A mere twenty-one years after the first war, World War II started with Germany's declared need for Lebensraum: room to live. It became the most devastating war in history, claiming 70–85 million deaths, about 3% of the entire world population of 2.3 billion at that time. The worst statistic of the war was the death of 129,000 to 226,000 Japanese people in Hiroshima and Nagasaki, which may forecast one of the possible disasters that could befall our planet, a nuclear war. The bombing of Japan took place in August,1945, ending the war. We have had three-fourths of a century to "get used to" that holocaust. The wrenching fear of the first years after the bombs were dropped seems to have faded.

It is appropriate to consider other aspects of the history of warfare and ask: What is in store for the future? The technology of warfare can be described in a rising curve: slow changes in weaponry and strategy for long periods of time, finally reaching a point where it moved upward faster and faster, soon approaching an exponential rate. We can therefore expect and fear even more sophisticated weaponry and tactics, such as cyber warfare and space warfare. Scientists believe that a fusion bomb explosion would be 500–1,000 times more powerful than the fission bombs dropped on Japan in 1945. Nine nations now possess nuclear weapons: United States, Russian Federation, China, India, Great Britain, France, Pakistan, North Korea, and Israel. There can be no question that that is nine too many. Such a weapon is a product of human genius and of human aggressiveness and stupidity, a frequently appearing combination of traits on our unfortunate planet.

War can also be a business. For example, in the early Middle Ages, during the Eighth to Eleventh Centuries, war was a transactional instrument for the Norse Vikings. They practiced both trading and raiding against the countries of Europe, for acquiring both goods and slaves. As a result of their conversion to Christianity during this period, there was also a minor religious aspect in their wars, as they undertook one of the early Crusades, in 1107, to wrest the Holy Lands from the Muslims who controlled the area.

Consider the latter days of sailing ships, when merchant convoys roamed the seas, carrying valuable cargo from various sources in the Far East to the seaports of European countries. They were subject to attack from pirate ships and privateers with intent to steal the cargo and sell the materials in what we would now call black markets. To forestall the thefts, the convoys often sailed under the protection of warships of one of the various European navies. Transport of goods by sea is still a major method of commerce.

After oil was discovered in the Middle East, the Western nations actively pursued obtaining control of the oilfields, beginning in 1908 when the British secured control over the first production in Iran. Oil was also found in several other countries, among which Saudi Arabia had the largest reserve and became one of the chief suppliers in the world. In mid-century, the suppliers nationalized their fields. A series of wars ensued at that time, starting with a conflict in 1948, when the newly created state of Israel was invaded by Egypt, Syria, and Jordan. There was a business component, a religious component, and territorial component as the bases for the wars, which constitute a danger to the present day.

Wars, ancient or modern, have been highly costly ventures and are almost always wasteful: of lives, body parts, and emotional stability, as well as bringing grief to whole families that lose members: fathers, mothers, brothers, sisters, sons, and daughters, all snatched away and gone forever. In addition, there has been massive destruction of various kinds of property. Tremendous amounts of money have flowed down the proverbial drain, although compared to the human losses, this wastefulness is almost trivial.

The term '*almost* always wasteful' used above refers to the inventions and discoveries made to prevent or respond to battlefield disasters, or logistical needs such as medical solutions for wounds or disease outbreaks, and new types of field equipment including walkie-talkies, duct tape, and radar. One could add nuclear weapons, which ended World War II, but 'tragically wasteful' is a much better descriptive term for that. There are about 13,000 war heads in existence. The mere existence of these weapons and their delivery systems, in the possession of nine countries, means that there is hardly a square foot of the earth's surface that is safe from a nuclear explosion or the collateral effects of intense radioactive fallout.

Regardless of the various ostensible reasons for countries engaging in war, some observers of the history of conflict tend to oversimplify the occurrence of wars by declaring that violence is bred in our bones. Others declare the opposite: that we are naturally gentle, which is also an oversimplification. My contention is that both groups are partially correct, and the expressions have a genetic basis. A major factor for making war is a predilection toward violence as opposed to gentleness in the human genome. This behavioral trait is governed by an unknown number of genes in the same fashion as other human characteristics. The effect of the genes is undoubtedly modified to varying degrees by various aspects of the environment. Greed, aggressiveness, intolerance, extreme ambition, and other behavioral traits also play parts in the propensity for war that exists among some people in this world.

These traits often act together to create a desire for power. If the number of detrimental alleles for these traits in a person's genome is relatively high, the desire for power will be strong. This sort of person might well crave a position of leadership in a war of conquest. Conquerors of the past such as Alexander of Macedon, Genghis Khan, Adolf Hitler, and Joseph Stalin could be included in such a group. Many of their followers may have had similar trait expressions but at a lower level of personal ambition.

Genetics and evolution have insured that we humans would be genetically different from each other, and therefore might participate in genetically based conflicts that range from name calling between two people to global warfare leading to monstrous losses through death and destruction. We know very little of the genetic makeup of the earliest humans, other than the generalization that they certainly had genes for the body pigments, height, internal organs, bone structure, intelligence, and behavior traits just as we do now. Many thousands of years later, we know that we have about twenty thousand protein coding genes leading to traits, plus many more non-coding genes that perform other tasks, such as turning coding genes on and off. However, we still know little about the actual genes of any one person of the eight billion people now on the planet.

Human ingenuity outdid itself in devising weapons for use in war. Thousands of years ago someone probably discovered that flailing with his arms, pushing, and hitting an opponent with his fist was not sufficiently frightening to get rid of an enemy. He discovered that a stick, a rock, or even a bone worked much better. One of those objects might well have been the first weapon.

The oldest devised weapons probably had sharp stone points, used as spears, dating back about 64,000 years. They could be used in close combat, or possibly thrown and used only one time. Then bows and arrows were invented, as well as swords, tridents, chariots, catapults, and flame throwers. Gunpowder was invented in the 10th Century and guns of various types were developed.

If we fast forward to the present and view a list of the various weapons that are available now and that have been used in, say, the last several hundred years, it provides a truly remarkable array. The list consists of many kinds of weapons, with variations: anti-aircraft guns, artillery pieces, biological weapons, chemical weapons, rifles, hand-to-hand weapons, siege weapons, bombs, and other explosive substances. They also include missiles, rockets, automatic and semi-automatic assault weapons, and nuclear devices.

It is perhaps indicative of the importance of war in the development of towns, cities, city-states, countries, and empires, that so much inventiveness has blossomed in the development of weapons over the many millennia of our sojourn on this earth. Also. it may accurately forecast a less than desirable future for us and our companion species on this beautiful planet. The inventiveness still exists, and we can anticipate weapons with still greater ability to kill, maim, and destroy.

Weapons development historically has followed a path that seems logical, a progressive path from small, simple beginnings to larger, more powerful, and more sophisticated technology: bigger and better guns, bombs, vehicles, and aircraft that are more damaging to the targets. However, there is one group of weapons that did not follow this path at all. Why? Because nature led the way, and we co-opted its handiwork. To wit:

Disease pandemics have swept over the world's populations episodically since the first recorded one: The Plague of Athens, in the years 430–426 BCE. was an attack of typhoid fever, which severely hampered Athenian troops in the Peloponnesian War and enabled the Spartan triumph. This was followed by the Antonine Plague (165–180 CE) (smallpox or measles) and the Plague of Cyprian (251–266 CE) (unknown cause). The Plague of Justinian was the first bubonic plague, recurring several times during the period 541–750. It is an infection of the lymphatic system usually transmitted by fleas, parasites of rats. The Second Bubonic Plague, known as the Black Death (1347–1353), caused more fatalities than any other pandemic in history. Thirty to fifty percent of the populations of various countries died. The onslaught resulted in a substantial world population reduction during the mostly steady increase since the early years of human habitation on earth. A Third Bubonic Plague first appeared in 1855, and is notable because a Swiss doctor, Alexandre Yersin discovered the causal organism, a bacterium named for him, *Yersinia pestis,* in the last decade of the 19th Century. The plague has re-appeared at intervals until 1994 in five states in India, but it caused only 700 infections. It may still be lurking somewhere.

Recent studies have shown that *Yersinia pestis,* and the Black Death, have been around much longer than we believed: The bacterium has been found in teeth of people in Britain and Germany who lived 4000 years ago.

Numerous other pandemics have invaded our world. They include cholera at intervals from 1817 to 1824 and then from 1961 to 1975; influenza from 1510 to 2010, including a major pandemic from 1918 to 1920; smallpox from the latter part of the 18th Century to mid-20th Century, when a successful vaccination campaign completely eradicated the disease, the only major human infectious disease to be eliminated. Pandemics that occur intermittently include measles, tuberculosis, leprosy, malaria, and yellow fever. At the time of this writing (2022–3), we were in a respiratory pandemic, caused by a new corona virus, SARS CoV-2. It is known as the COVID-19 pandemic, and its scope has been worldwide. Several vaccines have been developed and seem to be effective in protecting people who have been treated, including illness caused by two new variants named delta and omicron. Unfortunately, vaccination is not universally compulsory, and some people have refused to be vaccinated for various reasons, including the claim that it interferes with their freedom!

This sort of reasoning was more understandable when ignorance was the major guiding force, until we realize that the genes controlling personality and behavior are still with us and distributed in the same way as I have discussed earlier, from good to bad and everything in between on the bell-shaped curve. Ignorance still plays its part.

Infectious diseases also may become threats to human survival resulting from *our* actions, namely the use of disease organisms and viruses as weapons. There are three groups of disease agents, based on the targets of attack: anti-crop, anti-animal, and anti-human. Anti-crop and anti-animal agents are indirect attackers. They can be deployed to cause shortages of food for the civilian population in an enemy country, one aspect of so-called total warfare. Anti-human agents primarily target an enemy's armed forces.

Regarding anti-crop agents, articles in the literature discuss either the prospect of bioterrorist attacks on American agricultural plantings, or the development of anti-crop weapons by the United States. Two fungal disease organisms are the principal agents considered for use: *Pyricularia oryza* (rice blast) and *Puccinia graminis* f. sp. tritici (wheat stem rust). These organisms are meant to destroy rice and wheat crops, respectively, which are major food sources in Asia, Western Europe, and the Americas.

Herbicidal agents came to prominence during the Viet Nam war, when U.S. warplanes sprayed the jungle with materials such as Agent Orange, to destroy the foliage so that the North Vietnamese soldiers could be seen and attacked. However, in addition to the herbicide, the spray material contained highly toxic contaminants called dioxins. These substances affect humans, and cause problems with reproduction, development of body parts, and cancer. The effects may have also carried over to subsequent generations. American soldiers as well as the native peoples were apparently affected.

There are three principal anti-livestock agents. Two are viruses, one that causes rinderpest, a highly lethal disease of cattle; a second causes African swine fever. A third disease is psittacosis, caused by a bacterium, *Chlamidia psittaci,* to be used against domestic fowl species.

Finally, a rather large array of anti-personnel materials, consisting of viruses, bacteria, and fungi, plus toxins that are produced by some agents, can function as either lethal or incapacitating agents. The organisms and their diseases include the following: *Bacillus anthracis* (anthrax), *Brucella* spp., (brucellosis), *Francisella tularensis* (tularemia), *Vibrio cholerae* (cholera), Ebola virus, encephalitis virus, yellow fever virus, and *Rickettsia* spp., vectors which cause spotted fevers, typhus and other diseases. Possible toxins include ricin, botulinum, and fungal mycotoxins.

In recent years, the use of guns for non-military purposes has increased drastically, especially in the United States. It should be recognized that the founding fathers specifically and firmly declared that the bearing of arms was for the purpose of forming militias to protect the states from threats and attacks by the national government and not for individual use for any other purpose. That is the **only** reason the Second Amendment was written. The claims by the National Rifle Association, originally a gun safety group, emphasizing personal use for self-defense, was quite obviously not in the minds of the composers of the amendment. The consequence has been the murder of more people in the U.S. than in any other advanced country, with handguns and semi-automatic rifles specifically used to shoot and kill human beings.

In the following essays, I will discuss additional activities and major societal institutions, and their past and future contributions to the possible threats to survival of our species and our world. I don't pretend to be an expert compared to my experience in genetics and some related subjects, but I hope my thoughts on these subjects will be relevant.

3

Climate Change

The Brazilian rain forest is the largest forest carbon sink on earth. Its loss and replacement with crops, mining operations, and savannah would be catastrophic. Yet, the former president of Brazil, Jair Bolsonaro, has partially done away with protection given by the forest and has invited farmers, ranchers, miners, and lumber companies to cut much of it down. Similar damage is being done to other world forests, in Central Africa, Papua New Guinea, Southern Asia, and elsewhere. Fortunately, Bolsonaro lost his bid for re-election, and the Brazilian forest became safe, for now.

Human influenced global warming occurring now differs from the other threats to our planet in ways that have made it more difficult to deal with. Climate change is one consequence of the development of our world's civilizations that have provided us with numerous benefits, but also with non-beneficial results that threaten us in various ways. We occupy ourselves constantly with enjoying the beneficial results and fending off the bad results. Climate change has been warming our earth-home for many years, but most people have been unaware or dismissive of the changes that have been occurring. The changes have not been good ones, and the projected future effects will be worse. Despite warnings from those who have been studying the science of global warming-climate change, people have found it too easy to shrug off the warnings, and to believe that any serious consequences will happen to future generations. So why worry? Considering the various levels of awareness and concern among humans, what is in store for us and our home? The answers lie in the choices we make *now* for our future, especially for the near future, the next 20 to 50 years. Climate scientists have drawn up scenarios based on projected changes in greenhouse gas inputs to the atmosphere. One of them, formulated by the International Panel on Climate Change in Paris, projects four models: stringent mitigation, two intermediate mitigation scenarios, and essentially no mitigation. These are projected up to the year 2100. The first model predicts a greenhouse gas level essentially the same as in the year 2000, the intermediate ones will be almost twice that of the year 2000, and the fourth one over three times the 2000 level.

Similar changes can be expected for sea level rise, up to 2.5 meters increase for the least mitigating scenario, and for a continued rise in surface temperatures. In addition, droughts, heat waves, hurricanes, forest fires, and other forms and consequences of extreme weather will occur at greater frequency. When I first wrote this, it was August, and I was covered with perspiration. This may not seem unusual to most readers in the Northern Hemisphere. In Salinas, California, it is quite unusual. The Salinas Valley has been called the Salad Bowl of the World because it has a climate that features cool summer temperatures, ideal for growing lettuce and other cool season crops. Hot weather during the growing season has been a rare, and unwanted phenomenon. Many people who live here dislike hot weather as much as the lettuce does. Unfortunately, the climate is changing; extreme weather events are a signature of global warming. The same types of events are occurring in many parts of the world. They include severe storms, extreme heat, increased tropical cyclone activity, greater or lesser drought conditions, and increased or decreased river flooding depending upon location.

In the distant past, climate change on Earth was a natural phenomenon, probably initiated about 4.54 billion years ago when the planet was formed from stellar debris. The natural phenomenon continued until very recently, when humans unknowingly took over the primary task at the beginning of the Industrial Revolution.

The forces hypothetically unleashed in those earlier times were monumental. First, a wayward planet called Theia collided with Earth and knocked off a big piece that became our moon, which found an orbit averaging about 382,500 kilometers from Earth. The Earth became molten from constant volcanic activity, starting about 3.8 billion years ago. The planet finally cooled, developed a solid crust, and liquid water appeared. Tectonic plates were formed, becoming movable continents of different sizes and shapes, which varied in number from a single continent to the seven we have now.

The first organisms emerged about 4 billion years ago. Photosynthesizing life forms followed, about 3.5 billion years ago. They produced oxygen, which was released into the atmosphere, enabling the eventual development of multicellular organisms and the diversification and expansion of life. Arthropods (insects, spiders, crabs, etc. that have exoskeletons) came on the scene, followed by fishes and the earliest land animals, including dinosaurs and mammals, and eventually the forerunners of nearly all the phyla existing today. Recognizable members of the Genus *Homo* joined the population of living things about two million years ago, and *Homo sapiens* came along about 300,000 years ago. Early humans apparently formed small groups that lived as hunter-gatherers, until about 10,000 years ago, when agriculture was developed, and humans became relatively sedentary. Villages, towns, and cities arose and occupied the Fertile Crescent of the Middle East, followed by similar developments in other parts of the world, including China, India, Africa, and the American continents.

Change filled the air, figuratively and literally, with the beginnings of the Industrial Revolution in Great Britain, during the half century spanning the 18th and 19th Centuries. Starting with the textile industry, which had previously consisted of small family cottage businesses, industrialization meant that the activities of textile production were carried out on spinning and weaving machines dependent upon steel mills, railroad and steamboat shipments, and other mechanized procedures, all fueled by coal. The air in industrial locations was filled with coal dust and what we now call greenhouse gases. The dust was a visible pollutant in the industrial towns. As industrialization spread into Europe, the United States, and elsewhere, petroleum as a fuel for travel and other energy needs joined coal as increasingly potent producers of greenhouse gases, leading to global warming of the atmosphere and a mixture of climate change manifestations, not only in the air, but also in the soil and water that make up the surface of the earth.

Evidence for the human induced changes from production of greenhouse gases that affect the world's climate patterns is undeniable, except in the minds of those who either refuse to accept the evidence, or simply call it a hoax, because they fear losing the short-term benefit of making money in the energy industries. The deniers are primarily the leaders of those industries, plus dependent politicians, and members of the media who support them. The latter groups, in turn, persuade many in the naïve ranks of the general population that denial is recognition of "truth".

Until the industrial revolution, the climatic changes on Earth had been products of the Glacial-Interglacial Cycles during the Late Quaternary, each cycle lasting about 100,000 years. These could be called leisurely cycles, considering the time frames, as compared to the relative breakneck pace of the present global warming period.

What then is happening to our planet as the result of global warming, and what will happen in the future? Of the possible outcomes, what is the likelihood of human actions affecting those outcomes?

Human activity has caused the production of greenhouse gases: carbon dioxide (CO_2), methane (CH_4), and nitrous oxide (NO), which absorb radiant heat that would otherwise rise through the atmosphere and out into space. The average temperature has risen about .85° C. in the period 1880 to 2012. Most of the heat is being stored in the oceans and is melting the glacial and polar ice. The atmosphere, the oceans, and the continents are all increasing in temperature. Other symptoms include a rising sea level, increased humidity, greater storm activity, and more frequent extreme weather events.

There are other human actions that are contributing to the changes in earth's environment and climate. These include deforestation, plus the planting of crops such as cereals that do not have the CO_2 sink capacity of trees.

Instead, the use of chemical fertilizers and pesticides fills the air and waters with pollutants. Use of diesel engines leads to the deposition of soot. Soot settles on reflective ice, turning it into a heat absorber, and the ice melts, thus increasing the warming effect. Forests are major storage sinks for carbon dioxide and their loss adds to the warming effect.

The extensive data collected by climate scientists consist of two parts: evidence obtained from the geophysical records from many thousands of years ago, and actual recording of temperatures, water levels, glacier movement, greenhouse gas volumes, forest destruction, and other data from the contemporary years of the Anthropocene (human dominated period). This has enabled scientists to project likely further changes in climate and its effects under various scenarios. Some of these projections have been described above.

It is possible to carry the projections further into the future, enabling us to predict whether we can return to a cycle resembling those of the past or plunge ourselves into a "hot earth" trajectory in which the earth may have difficulty supporting human life as well as the existence of many of our contemporary species.

The rate of CO_2 build-up at present means that we have already passed the point of no return, so that temperature will continue upward, the CO_2 level will rise further, and the sea level will continue to rise, regardless of decisions we make. The changes we must make are toward maximum mitigation actions, beyond the goal of the Paris political agreement, to prevent still higher levels occurring.

The lower mitigation paths will lead to a "hothouse" earth. This is unknown territory and should be frightening even though it will probably not happen during the lifetimes of anyone living now. Those who have just been born and those born in the future will be increasingly likely to suffer the consequences of our neglect. The actions that we take now will lead to bio-geophysical feedback, which connect biological activity with geological and physical phenomena. Active steps to reduce the greenhouse gas buildup will lead to negative feedback, which in turn will minimize the extent and consequences of sea level rise, glacial loss, violent weather patterns, destruction of carbon sinks, and other bad effects. Failure to take sufficient action will trigger positive feedback that will worsen the consequences: continued loss of glacial ice, rise in sea levels, thawing of the tundra, loss of forest sinks, and on and on.

As it has for 4.5 billion years, the earth itself will follow its own continuing path of climate change into the future, however long lasting that future might be, facing all events as they arise, unaware of consequences, catastrophic or not. Non-human inhabitants will be, as always, barely aware of the events taking place, but will feel the consequences. Some species may be more aware of the changes occurring. Only our species will be cognizant of and increasingly horrified by the course of events: the basic events of nature embellished by the actions of humanity. Even the deniers may someday be finally forced to admit that humanity must assume major responsibility for global warming. In doing so, they may realize that the physical changes taking place also have mostly bad biological consequences, affecting all creatures on earth. Will all this continue to happen until we reach the point of no return?

Our principal concern is, naturally, with the effects on us, *Homo sapiens*. We should keep in mind, however, that all the other forms of life on earth have been innocent bystanders of environmental change from their first appearance on the planet over two billion years ago, experiencing all the evolutionary changes since then. All the evolutionary paths since the beginnings of human induced climate change will be modified and speeded up as the climate continues its warming way, especially if we persist in ramping up greenhouse gas production by employing inadequate measures of control.

Global warming will affect humanity directly by impacting health, the environment, social norms, living conditions, and other aspects of our lives. A major effect will be on our ability to produce food and other agricultural products. The onus will be on farmers to adapt their crops to the higher temperatures, and plant and animal breeders to develop new cultivars and breeds adapted to drastically different living conditions.

The danger to health will include a broad array of direct and indirect effects, with increasing impact if our mitigation efforts are insufficient. These include an increased spread of insect borne diseases as rising waters and temperatures force insect habitats into new and larger territories and support the growth of larger populations. Air pollution rates will increase. Flooding, extreme temperatures, and frequency of droughts will increase, along with greater danger of wildfires.

This situation is one of many where the brunt of the problems will fall in areas and populations that are less able to cope with the climate changes, due to poverty and location, than in less vulnerable areas. They will include small island communities, river deltas, and other coastal areas, especially mega-cities to which poor people, the elderly, and young people have been migrating in droves, particularly in less developed countries. Therefore, those people, who already lack many of the benefits of civilization, will suffer greater impact than people in more benevolent environments.

It is likely that climate change will affect all or nearly all human activities, lifestyles, etc. Therefore. the most serious question about our future is: What path will we follow to confront the developing threat: strong, moderate, or negligible mitigation? I am very much afraid that, worldwide, we will take the moderate or minimum remedial steps. Powerful political and corporate figures have made it convincingly clear that in the most important countries of carbon neglect they will do little or nothing to alter the present course. Denial and greed make a powerful statement. The unstated excuse is that they will not be around to see the results anyway, so I repeat: they don't care what happens in the years to come.

Environmental pollution and global warming-climate change are cousins. Both affect the same components of the environment: the air, water, and soils plus the senses, health, and welfare of the earth's living populations. Climate change is a more insidious invader and unfortunately is more easily denied by those who worry that their short-term current income and wealth are at stake, and others who are politically beholden to monied groups. This situation is further explored in the essay on inequality.

The climate control agreements signed in Kyoto, Japan in 1997 and in Paris in 2016 are considered by many climate scientists to be doomed to failure for several reasons. One is that neither one is legally binding, as there are no penalties for failure to comply. Another is that the United States, one of the main culprits in the production of greenhouse gases, which cause increased temperatures, twice withdrew from both agreements by order of President George W. Bush and later President Donald Trump, to protect American energy suppliers from short-term profit losses. Fortunately, President Joe Biden, on his first day in office, signed an executive order to re-admit the United States to the Paris Agreement. Other climate culprits, especially Brazil, have also neglected the problems. In addition, the consensus among climate scientists is that the goal of a 2^0 C rise over pre-industrial times should be 1.5^0 C, a more meaningful goal. The prospects for the future are not good.

Critics and nay-sayers have added another argument to their denial of the dire consequences from global warming and climate change. Republican Congressman Lamar Smith, former chair of the House Committee on Science, Space, and Technology, stressed the benefits of higher concentrations of carbon dioxide in the atmosphere. It combines with water (H_2O) to initiate photosynthesis, which produces carbohydrate, a foodstuff that is necessary for plant growth. Therefore, the larger amount contributes to increased plant growth and yield of edible and otherwise useful plant products, such as seeds, fruit, and leaves. However, other research has shown that the nutrient value of staple crops, like wheat, barley, and rice, is reduced at high levels of CO_2, requiring the addition of iron, zinc, and protein to compensate for the loss.

In other words, the beneficial effect of small increases of CO_2 will be more than overcome by continued increase in the atmospheric content of the gas. Latitudinal areas now functioning as the superior areas for growing crops will lose that advantage to areas to the north or south. Crops will need to be moved to the new production areas and will also need to be bred to produce genotypes adapted to those areas.

In addition, fruit and nut trees will benefit less than row crops with the warmer temperatures. Those tree species require a winter chilling period to encourage the development of flower buds. Another consequence will likely be unreliable rainfall seasons. To sum up, increased CO_2 in the atmosphere will have both positive and negative effects on plants grown for food.

The top ten countries attacking climate change in 2022, albeit from different approaches, are: 1. Luxembourg, 2. Denmark, 3. Morocco, 4. Netherlands, 5. Lithuania, 6. Portugal, 7. France, 8. Finland, 9. Sweden, 10. China. This ranking changes from year to year. Also, no country merits a very high rating, since they fail to follow a path leading to the goal of 1.5^{0}C limit, which requires stringent mitigation.

4

Environmental Pollution

Now let's turn to environmental pollution. This phenomenon began with the earth's own disturbances: volcanic eruptions and forest fires. Of course, on the ancient earth, with no living beings, pollution hardly qualified as a harmful disaster. Millions of years later, however, the industrial revolution made it a serious worldwide problem. In the mid and late 19th Century, the principal sources of pollution were the burning of coal, industrial chemical discharge, and sewage disposal. Over the years, human minds have contributed a tremendous number of inventions towards the world's economic progress, solving problems in every human endeavor. Unfortunately, in many of these activities, various affronts to the environment were part of the results. Now the totality of environmentally offensive contaminants seems a magnificent monstrosity. I don't think this is an overstatement.

Let's consider the evidence: the forms of pollution that we must confront. Like climate change, they are problems that require extrapolation into the future, which may be denied by some, because they are not yet fully manifested. Forms of pollution are evident now and are obvious and usually annoying and have been doing harm for years, from sources that are human creations, such as manufacturing, transportation, extraction from the planet's surface, poor waste disposal, forms of transportation, and agriculture.

Air pollution is the most familiar form; it has been a problem for many years and usually is visible. It consists of gases and fine dust, mostly produced by motor vehicles, industrial waste, and agricultural chemical sprays, delivered by airplanes, helicopters, or tractors. Gases include ammonia, carbon monoxide, sulfur dioxide, and the global warming gases such as methane and nitrous oxide. Also included are organic and inorganic particulates. These materials may cause diseases, allergies and even death to humans, animals, and crops. Mounting evidence suggests that air pollution can cause intelligence and psychiatric problems. Outdoor air pollution from fossil fuel use causes about 3.61 million human deaths annually. It is considered the world's highest environmental health risk. About 90% of the world's population breathes dirty air at some level.

People exposed to high levels of certain air pollutants may suffer from irritation of the eyes, nose, and throat, wheezing, coughing, chest tightness, difficulty breathing, worsening of lung and heart disorders like asthma, and increased susceptibility to heart attacks, cancer, and even death. However, there are control strategies possible that are effective if they are used. These strategies are not universally used, unfortunately. Also, possible environmental effects include acid rain, accelerated eutrophication, haze, and health problems of wildlife. It may also cause ozone depletion, and crop and forest damage. Last but not least, is global climate change.

Light pollution is produced over large cities at night. Groups of cities in several small European countries as, for example, Belgium, appear as a single mass of light when viewed at night from above. Similar effects occur in the big cities in the U.S.A. On the ground, one cannot look at the stars and planets through a telescope because of nearness of the background light, which obscures celestial objects. There are several other types of light pollution. Light trespass occurs when yard lights on a property shine over the fence into a neighbor's yard. Other light sources can also be annoying: over-illumination, glare, light clutter, and light from satellites; the names are self-explanatory.

Another form of pollution is litter, discarded man-made materials found on public and private properties. Litter has a history. A road marker in ancient Greece stated: "whoever drops their litter on the street owes 51 drachmae to whoever wishes to claim them". Nevertheless, litter has remained a problem to this day. Auto tires are the most obvious dumped form of litter and can harbor insect vectors that transmit human diseases. Containers such as paper cups, plastic bottles, and aluminum drink containers may fill with rainwater and harbor mosquitoes. Cigarette butts are a threat to wildlife and can also start fires. The list goes on.

Governments at different levels have instituted laws designed to deal with the problem of litter. Their responses to the problem include litter bins, from which the materials can be removed to be reused or recycled. Cleanup events may be scheduled by local authorities. Earth Day cleanups have been held globally since 1970. Commercial properties often have litter picking maintenance programs. All in all, these and other cleanup actions form a constant response to the many ways that litter is created. This is a problem that may never be solved permanently.

Noise pollution emanates from roadway vehicles, aircraft, and industrial activities. In the oceans, high intensity sonar, used primarily by submarines for navigation, also adversely affects communication among cetaceans and can also harm their nervous systems.

One of the most offensive types of pollution is manifested by plastic products, including microplastics, the small pieces remaining from deterioration over time. On land some of the materials are collected at dumping places.

Collection and disposal procedures for recycling vary considerably within and among nations. For plastic and other forms of waste, collection level ranges from none to over 55% in Germany and nearly 100% in Sweden. The collected items include newspapers, mixed paper, cardboard, magazines, plastic bottles, aluminum and steel cans, and glass containers. Single use plastics are difficult to deal with and often wind up in the oceans or in other waterways. In the oceans they may form large floating islands resulting from rotary water movement or gyres. Other plastics also often wind up in the water.

Soil contamination consists of chemicals from spills or underground leakage, and from application of fertilizers and pesticides. Water pollution occurs due to a combination of several causes: discharge of wastewater, sewage, and urban runoff, plus agricultural runoff, which often contains harmful chemicals.

Electromagnetic pollution stems from a diverse range of sources, some of which may affect human health. These occur at low frequency, mostly from power lines and electrical appliances.

Radioactive contamination is a product of nuclear

power generation and nuclear weapons research, manufacture, and deployment. Nuclear warfare has not burst out since August of 1945, and yet it remains one of the greatest potential threats to our survival. If nuclear weapons are used in the future, we can expect extensive radiation coverage of large areas of the earth's surface, causing death, injury, and genetic damage to many thousands or even millions of people worldwide.

The various forms of environmental pollution are unsightly, physically annoying, and often toxic for nearby residential areas but less likely to be of widespread danger to large populations. Radioactivity, however, can be dangerous in large areas if it is a product of nuclear explosions.

As with global warming, there are often deliberate corporate and political efforts to deny the existence of environmental pollution and its dangerous effects, attempting to prevent action to control their dispersal and therefore endangering present and future populations. Sometimes legal action is necessary to prevent the dangerous actions. Constant monitoring and efforts to control the problems are necessary to minimize dangerous and/or unpleasant effects.

How does environmental pollution qualify as a threat to survival on Planet Earth? As presently manifested, the various forms of pollution range from mildly annoying among some population groups to a health threat affecting a substantial number of people, especially those victimized by poverty and other life deprivations. The degree of danger, however, is not comparable to war or to climate change, both of which can wreak havoc on an immense scale. War is immediately deadly, while climate change becomes increasingly harmful over long periods of time. If it reaches the "hothouse" earth stage, there may be no aftermath of recovery.

5

Excess Population

War and modern-day climate change are two of humanity's dangerous creations that threaten the survival and well-being of all biological species on Earth. Additionally, several other human advances can increase the impact of those dangerous creations. Excess population size is one example. It can be controlled, if we choose to do so, despite the religious strictures and natural desires that have largely prevented that from happening.

The human population (*Homo sapiens)* appeared about 300 thousand years ago on the African continent. The United Nations has estimated the size of the population, beginning at 70,000 BCE, after an event labeled the Toba Catastrophe, an enormous volcanic eruption in Indonesia, which led to a severe reduction in global atmospheric temperatures. The remaining population was estimated at 1500, a reduction from about 20,000 before the event. By 10,000 BCE, the population increased to about 4 million. The number grew relatively slowly but steadily, except for setbacks attributed to the plague and other diseases, and reached its first billion about 1800, at which time exponential growth began. By 2023 about 8 billion people were recorded. Our numbers now are continuing to rise, but more slowly. We may reach 10 billion persons by mid-21st Century, on the assumption that a world catastrophe does not occur first.

Early humans existed for thousands of years in small hunter-gatherer groups until the invention and development of agriculture approximately 10 thousand years ago, which led to a more sedentary life for people, who became farmers and residents of villages, towns, and cities. It is probable that this life change marked the beginning of many human endeavors that have brought us to the modern world. It offered the opportunity and the time for people to use their minds to develop ideas and methods, enabling them to improve their lives. Unfortunately, their new lives also provided the means to do harm to each other and to the planet.

As the population grew, people migrated to the virtual four corners of the world, exploring, and then settling on every continent and in all climates, on the deserts, mountains, and islands, from the tropics to the arctic lands. Even Antarctica has a population, about 5000 research people during the summer, which shrinks to about 1000 during the winter.

As previously noted, significant interruptions to the overall growth path occurred several times in the past, when worldwide disease pandemics took the lives of millions of people. The Plague of Justinian was the first known pandemic, a bubonic plague occurring in the years 541–549 CE, taking a toll of 50% of Europe's people. The plague returned in the 14th Century (1347–1351). It was a dreadful onslaught of disaster in Europe and Asia and was branded as the Black Death. It overcame the upward birth trend and turned the population curve down for years until completely fading away. Milder versions of the plague returned several times after that. During the late 19th Century, its cause was finally identified as a bacterium, *Yersinia pestis*, named for its co-discoverer, Alexandre Yersin, a Swiss bacteriologist. Eventually, treatments were discovered, and the disease is now a relatively minor problem. Hopefully, it will remain so.

The historic first appearance of the plague happened earlier in time than we had thought, with the recent discovery of *Yersinia pestis* in human teeth in Asia and Europe dating back to 2800–5000 years ago.

Overall, the population has continued to increase despite the setbacks and has become another force that threatens the stability of the planet. There is no doubt that the number of people on the surface of this earth will grow even more, barring a monstrous killing catastrophe. Aside from the actual numbers, the time that we reach certain thresholds, the consequences from reaching those thresholds, and the possibility that some or all the consequences will be fatal or near fatal to us and our planet must be foremost in our minds.

Early concerns about overpopulation were stated by those whose knowledge was lessened by an inability to see beyond a very limited horizon. Tertullian, 155–220 CE, a resident of the Roman province of Carthage, declared that "What most frequently meets our view…is our teeming population… burdensome to the world, which can hardly support us…" He touted pestilence, famine, and war as *valuable* means of pruning the population. Fifteen centuries later, in 1798, Thomas Malthus published an essay stating that the population was increasing at a geometric rate, while the food supply was increasing in an arithmetic progression. He wrote that eventually the population would outgrow the available food supply, and mass starvation would result.

Malthusian followers, then and now, also blamed this prospective future on the poor, who were prone to have 'too many' children. As in so many debates over human issues, this explanation is greatly over-simplified. Crowded populations will continue to exist in some parts of the world, heavily dependent upon the availability of often scarce basic resources. Goods and services will probably be available in many urban areas, regardless of the number of inhabitants. That's where the money is! In marginal environments, unfortunately, these benefits are likely to be lacking. Water and/or arable land will be scarce in the desert, in many mountainous areas, and in other remote locations.

It may be that the greater deleterious effects of large populations are those that damage the air, rivers, land, and oceans. More people will produce more pollution, increasing the damage to the health of humans and other creatures. There will be increased greenhouse gas production, continuing the rise of global temperatures, melting the glaciers, and raising sea levels worldwide. There will likely be negative reactions in crops, possibly reducing yields and affecting food supplies; and possibly even causing wars over the control of resources.

Suppose that we wish to temper population growth on the assumption that such an action is or will become a desirable practice. How is this to be done? We can ignore the niceties of our civilized ways and institute various eugenics measures to reduce births, like those practiced by Nazi Germany, the United States, and other countries in the recent past. In 1980, the government of China set a mandatory one-child-per-family policy, except in rural farming areas, where two were permitted. This edict has been changed twice recently: In 2016, married couples were allowed two children, and in 2021, this was increased to three. They are no longer forced to apply for a family planning service certificate.

There is still criticism for having any restriction at all. For example, Huang Wenzhen, a demographic expert in Beijing, claims there should never have been any restriction. He says that the policy should be fully liberalized, and that giving birth should be strongly encouraged. However, even now, the Uyghurs and other Muslim groups are still being forced to have fewer children to suppress their population growth.

Even worse, forced sterilization of "undesirables", based on intelligence, color, religion, or other wicked reasons is another practice that may be accompanied by a matching policy of encouraging an increased birthrate of "superior" people.

It is not unlikely that the more benign and fruitful means of attacking the possible deleterious results of population increase is to remedy the secondary aspects. The question may be posed as follows: Can we do what is necessary to reduce the enormous human footprint, especially by reducing or eliminating the following practices: consumption of meat, use of fossil fuels, destruction of forests, and abuse of technology? We must also reduce pollution, elect responsible governments, improve our public education systems, expand our knowledge of science, improve business practices, improve agricultural methods, avoid war, and overcome our genotypical difficulties, including ethnic, racial, gender, political, and religious prejudices. This is a tall order.

During the decade from 2011–2020, twenty-three countries have had a net decrease in population. The countries include fifteen from the Balkans and Eastern Europe, four from Western Europe, and four from other parts of the world. The losses are based primarily on net changes in population events: emigration rates exceeded immigration rates, and death rates exceeded birth rates.

So, we have questions that may only be partially answered. We may know the real effect of a population increase sometime in the future. If the rate of growth slows down sufficiently, stabilizing at a level of about ten billion, will that allow us to live relatively comfortably with each other? Will we be left with one less threat to deal with? Unfortunately, the real population threats listed in the above paragraphs will still be there unless they are dealt with as well. It is necessary to keep in mind that whatever happens to the population of the world, many people are already victims of various other inequalities. The population size that we reach will adversely affect more people. We must help solve their problems also.

Another concern that requires consideration is the effect of our population growth on the populations of other species, such as deer, wolves, birds, ocean mammals and fish, and many plant species. Since 1970, the number of wildlife populations has plunged 69%. Not a single species has escaped the downward list. A total of over 105 thousand people is under some level of threat of extinction. Affected by these changes is also the maintenance of biodiversity as well as the sheer number of species lost or on the brink of disappearance.

An important factor contributing to the losses has been the destruction of the world's great forests, in Brazil, New Guinea, and elsewhere. Large acreages have been destroyed to permit ranching, mining, and other money-making activities. Those actions have contributed considerably to global warming and climate change, as described elsewhere in this volume.

A report produced by a world organization, the Intergovernmental Science-Policy Platform on Biodiversity and Ecosystem Services, warned in 2019 that the rate of species extinctions is accelerating, with serious threats to people and other species around the world: about 1,000,000 species are in danger of extinction. The average abundance of land-based species has fallen about 20%, amphibian species about 40%, and about 33% each of marine mammals and reef forming corals. At least 680 vertebrate species have become extinct since the 16th Century. Where does the responsibility lie? There are five major blameworthy changes for the above numbers: (1) changes in land and sea use, (2) direct exploitation of organisms, (3) climate change, (4) pollution of various types, and (5) invasive alien species.

Those changes are the results of actions taken to satisfy the desires of the prosperous segment of the human population. These actions largely ignored the needs of the many: including the poor, the helpless, and residents of rural communities, who live on the outskirts of society. Rapid declines in biodiversity, support of ecosystems, and the ability of nature to support the needs of the world population means that the societal and environmental goals will not be reached because of the current methods and likely future trajectories in the selected directions. Not only are indigenous populations being ignored under our present policies, but they are also suffering with their loss of control over the environmental products that contribute to the prosperity of the well-to-do, who usually benefit from the acquisition of products from indigent locations.

The report presents the necessary goals and actions to reach those goals in the various systems of agriculture, marine and freshwater systems, and urban areas. Each system requires transforming actions to enable all of us to reach the necessary goals of sustainability and equalization of the various benefits.

It is essential that we promote the proper practices to change direction and act to reach the various necessary goals in the different areas of the planet suffering from our neglect, in the following ways:

Agriculture—Promote good agricultural practices, proper landscape planning, greater engagement of all participants from the producers to the consumers, integrated landscape and watershed management, genetic diversity, reformed supply chains, and reduced food waste.

Marine systems—Improve fisheries management: set sensible quotas, protect needy areas, manage biodiversity areas cooperatively, and reduce run-off pollution.

Freshwater systems- Collaborate in water resource management and water use reduction, end soil erosion and pollution runoff, and increase water storage.

Urban areas—Provide healthy urban environments and access to green spaces for low-income areas.

Overall—Support a variety of value systems, and welcome participation of indigenous people and local communities. If we can take up these societal improvements, such actions should contribute to population growth reduction.

6

Science

Science is the search for and acquisition of knowledge. That statement would seem to absolve scientific activity of blame as a threat to survival. Unfortunately, new information may be used in further research and development in applied sciences, engineering, and technology. The products of this second stage might be useful in various ways. Presumably most would have benevolent uses, while others might be useful in war, help produce the deleterious products of pollution, contribute to global warming, and/or lead to inequality.

An example of this sort of double life is a material invented in the early 9th century in China by alchemists in search of an elixir of life, a material expected to confer immortality. That outcome did not occur. Instead, the mixture of saltpeter (potassium nitrate), sulfur, and charcoal turned out to have explosive properties, and was eventually designated as gunpowder. Few inventions have had an equivalent negative impact on the history of the world.

Gunpowder is used in weapons of warfare and in other ways that cause death, destruction, and other sorts of violence. That usage more than balances its useful properties, such as self-defense, construction procedures, hunting, production of fertilizers, and mining.

On the other end of the scale, consider Alexander Fleming. In 1928, he stacked a group of Petri dishes with cultures of the bacterium *Staphylococcus aureus* on a corner of his laboratory bench before leaving for a beach holiday. Upon his return, he found one culture contaminated by a fungus, *Penicillium* spp., which was killing that part of the bacterial colony that it touched. He gave the name penicillin to the active substance produced by the fungus. Penicillin became one of the early antibiotic medicines and was used extensively to treat wounds during World War II. It was difficult to produce large enough amounts of the original drug, and further research led to the development of several similar substances that could be produced in sufficient amounts. Consequently, they are still in use.

Two of the most important scientific discoveries ever made were those by Charles Darwin and Gregor Mendel, published in 1859 and 1865, respectively, but unknown to each other. Mendel's discovery led to the development of the enormous field of genetics with its myriad array of sub-sciences. Strangely enough, it took a rediscovery to alert scientists that Mendel's data, when studying the inheritance of plant characters due to single genes, was correct. His experiment showed the inheritance of single genes controlling characters of garden peas: flower color, seed color, seed shape, and plant height. The results were published in an obscure journal and thus were essentially hidden from the scientific world. However, in 1900, Hugo De Vries and Carl Correns separately duplicated Mendel's work and credited Mendel with the original results.

Darwin's theory of natural selection (also published independently by Alfred Russell Wallace) challenged an entrenched religious belief system about the origin of life on earth, which led to a controversy that still exists.

A famous trial illustrating the latter half of that statement took place in Dayton, Tennessee in the USA. In 1925, high school science teacher John Scopes talked a bit about evolution in his biological course. An uproar by Christian evangelists arose, claiming that Scopes violated the Butler Act, which forbid the teaching of evolution in Tennessee. Scopes agreed to be tried and was found guilty. The trial featured two of the best-known attorneys in the United States, William Jennings Bryan for the prosecution and Clarence Darrow for the defense. It was agreed by many observers at the proceedings and from many parts of the nation that the trial did not celebrate the law but in fact was a circus. Genetics and evolution together tell the real story of life on earth beginning many millions of years ago and carrying on until the present.

Albert Einstein, in the early part of the twentieth century, developed two closely related theories about the relationships among universal forces: The Special Theory of Relativity (1905) and the General Theory of Relativity (1916). The Special Theory explains the relationships of all laws of physics (space, mass, speed, and time) except gravity. The General Theory explains the relationship of the same forces and includes gravity. The theories themselves are very difficult to understand for nearly all people, except those physicists who have studied and been active in pursuing truths and applications derived from the theories.

Based upon these theories, Einstein derived the equation $E=mc^2$, which is commonly seen in various publications, non-scientific as well as scientific. Many people know that the symbols stand for the relationship between the amounts of energy(E), mass(m), and the speed of light(c) The last is a very large number: 299,338 kilometers per second. It translates to the fact that a relatively small amount of mass converts to an enormous amount of energy under certain conditions. When treated as a source of usable energy, the conversion process produces useful energy for benevolent purposes. When used as a bomb, the conversion is swift, becoming the most explosive device ever created, capable of causing tremendous death and destruction: In early August 1945, two atom bombs were dropped in succession by United States Army Air Corps B-29 bombers on the Japanese cities Hiroshima and Nagasaki. Both cities were destroyed, about a quarter of a million people were killed, and thousands of others died of radiation poisoning. The Japanese responded by surrendering, and World War II ended. Since that time, there has been no repeat of weaponized use of atomic power, due primarily to forbearance of the countries that have developed nuclear weapon capability. However, the threat lurks as a frightening prospect for our future.

Aside from war and weapons, nuclear energy has peaceful uses. Most familiar to us are the nuclear power plants that have been built around the world to supply electricity to large numbers of people. Less well known are radioisotopes, with applications in agriculture, consumer products, food irradiation, industrial procedures, and medicine.

Radioisotopes are also used for carbon dating of ancient fossil discoveries. The relative amounts of non-radioactive and radioactive isotopes are measured in rocks to determine their age and the age of fossil remains associated with the rocks. Irradiation technology is used in treatment of foods susceptible to spoilage: gamma rays are employed to kill bacteria that can cause food borne diseases. In industry, radioactive tracers are employed to detect leaks, inspect metal parts for defects, measure thickness of metals, and other uses. In the field of agriculture, mutation breeding refers to the process of creating mutants in plants that have potential usefulness for variety improvement, as for example mutant genes that confer disease resistance. Insect control can be enhanced by sterilization of captive males, which are then released in the field, so that mating produces no offspring, thus reducing the insect population and its ultimate damage to a crop. There are many other uses, in therapy, nuclear powered ships, generation of heat in space vehicles, and other applications.

Science may come into conflict with a belief system, stemming from the scientific method itself. Science is the search for truth, confronting an unknown cause of a phenomenon with the formation of a theory, based upon observation of certain facts that seem to be in some way related to each other. The theory asks the question: Are these facts really related? The investigator starts with a so-called null hypothesis, which states that they are *not* related, and then conducts an experiment to determine if the null hypothesis was correct. The conclusion reached depends upon the results, which show that the hypothesis was either right or wrong.

This procedure is not dependent upon a previous belief, which might well have been wrong. Galileo Galilei is one of the famous early scientists who encountered the force of an intrenched belief system. He defended the work of the mathematician Copernicus, which stated that the Earth was not the center of the Solar System and instead revolved around the sun. Galileo was condemned by the Catholic Church for opposing a sacred Aristotelian theory. He was placed under house arrest for the remainder of his life.

One strength of science is in its occasional perversity, shown by Galileo after the verdict against him. "Still, it moves." he murmured, referring to the earth's motion around the sun. The scientific method is rooted in this concept. The statement of a theory is followed by a continuing effort to prove the statement wrong. Even when further research efforts to test the truth of the theory seem to confirm its validity, there is always the thought in the back of the scientific mind that someday, the truth will out, showing the theory to be wrong after all. It may take a very long time. It also may not happen at all, preserving the truth of the original conclusion. Additional related experiments may also confirm the truth.

The two branches of biology that form the basis for this book also have aspects of controversy. Genetics has been misused in its application to eugenics. Evolution has been repeatedly denied to this day by many members of some religious belief systems because it may be in direct conflict with those beliefs. Those beliefs state that the universe was created in six days, and the first two people created were named Adam and Eve.

The desire to turn people into better versions of themselves, or to prevent persons labeled as undesirable from reproducing themselves, originated before anything was known about human genetics. In ancient Sparta, every newborn child was inspected by a council of elders that decided if the child was fit to live. In early Rome every father was directed to kill a child who was physically defective. Neither practice had any basis in truth. They were simply cruel. The English polymath Francis Galton based his eugenics theory of improvement of the human population on a statistical understanding of heredity. Eugenics measures have been used in Great Britain, the United States, Nazi Germany, and other countries. Such usage is cruel and inhuman, for which there is no excuse.

Ever since Charles Darwin and Alfred Russel Wallace independently published their theories of natural selection, the concept has been strongly denounced by some branches of Christianity and other religious and political belief systems. Nevertheless, evolution has been shown to be real, the basis for the existence of millions of life forms that have populated and adapted to the entire earth including the driest, wettest, lowest, and highest lands, the deepest oceans, the coldest and warmest areas. These extreme climate places became populated with a tremendous variety of creatures, of different sizes, forms, abilities, adaptabilities, and many other traits. All the similarities and differences on this relatively tiny planet exist because of evolution as fueled by genetics.

Nevertheless, evolution has been denounced by members of religious and creationist groups who believe that the biblical statements are undeniable truths: the universe was created in six days; people lived for hundreds of years; the entire world was inundated by a massive flood, and only one man and his family survived; Jesus was crucified, died, returned to life, and went to heaven; humans are unrelated to other creatures; and so forth. One outgrowth of this sort of belief system is the base canard that Jews were and *are* Christ-killers. Also, some teachers tell their students that humans and dinosaurs existed at the same time, a statement that is wrong by millions of years.

Science has evolved because of the cognitive ability of the human mind to think, compare, and develop ideas. The earliest protosciences on earth were astronomy and agriculture. It is quite easy to understand the attractions of the sun, the stars, the moon, and the planets, the strange movements of those objects in the sky, and their locations. For some, the phenomena of how and why also intruded in their minds and led to centuries of astronomic investigations.

Other roads led to the planting, growth, and cultivation of formerly foraged plants as actual crops, sparking the improvement of wild maize and potatoes in the Western Hemisphere, and cereals, peas, chickpeas, and flax in the Fertile Crescent of the Middle East. Domestication of sheep, cattle, and pigs was also part of the agricultural revolution. Dozens of crops originated in nine centers of diversity: Mexico and Central America, South America, North America, the Mediterranean basin, the Middle East, Ethiopia, Central Asia, India, and China. The Indian Center includes Indo-Burma and Siam-Malaya-Java as sub-centers.

Looking over all the contributions of the various sciences, it is difficult to label them as threats to survival in the same class as war and climate change. Of course, they are not. Whereas war and global warming are relentless forces of destruction moving in a single direction towards eventual worldwide disaster, there is always a choice of 'do or don't' in the sciences. Science is the search for truth. It is not a force. It is a procedure enabling the acquisition of knowledge, and we always have choices regarding that knowledge: whether to use it and how to use it.

There is the rub. Before the start of World War II, scientists in Germany, directed by their Nazi leaders, tried to develop weapons with far greater explosive power and technological advancement than those in use at that time. These turned out to be the V-1 and V-2 cruise missiles and the atomic bomb. The missiles were successfully developed and were used to bomb Great Britain in 1944–5. Fortunately, the nuclear weapon project had not reached fruition at the time that Germany surrendered to the Allies, ending the war in Europe. Two of the main reasons for the German scientists' failure may have been their inability to work cooperatively plus the non-availability of sufficient funding.

The United States, Great Britain, and Canada also began research in the late 1930's into the feasibility of creating nuclear explosives. Early in the War, when allied intelligence discovered the German atomic plan, it was decided to attempt to develop and construct atomic weapons. A letter was written to President Roosevelt in 1939, advising him that the Nazis would try to create an atomic weapon, and recommending that the Allies should also begin a weapons program. The letter was signed by two scientists: Albert Einstein and Leo Szilard. President Roosevelt agreed and initiated a program that became known as the Manhattan Project. Unlike the Nazi experience, the project was generously funded and more cooperatively run and was therefore successful in producing atomic weapons that were used to end the war in the Pacific.

In the end, the good guys were successful, and the bad guys were not. But considering the terrible damage done by the only two nuclear weapons used in a war, can we be truly pleased by the decisions and the outcomes? The controversy over the use of nuclear weapons on Japanese cities at such human cost continued for many years. One of the main justifications for their use was that they may have saved many thousands of lives that would have been lost if it had been necessary to conduct a massive invasion of the Japanese islands. That effort would have included American troops, and Japanese soldiers and civilians. The total losses might also have included Russian troops if the Soviet Union had chosen to participate in the invasion. That speculation is part of the unknown because the invasion did not take place.

Genetics, modified by environment, controls the

direction taken with the opportunities offered by scientific discovery. As with all other phenomena discussed here, the direction is dependent on the behavioral traits of the people involved, and on the distribution of alleles of appropriate traits in those persons interested in applying the genetic information for various purposes. Such purposes can be benevolent or evil.

The modern array of sciences has been developed over thousands of years, beginning in prehistoric times, before writing had been invented about 3400 BCE. The knowledge and the ways in which the information was used depended on its oral passage from each generation to the next. The earliest sciences, primitive agriculture and astronomy, functioned in this manner. Some basic information about human physiology was discovered in plant research, as well as information about the characteristics of plants and animals.

The accumulation of scientific knowledge began to develop in Sumer, a part of ancient Mesopotamia, about 4000 years ago, with studies in astronomy. Then followed centuries of manifold scientific discoveries all over the world. Early on there were advances in astronomy in Ancient Egypt. Later, studies were carried out in Ancient Greece, proposing very early that the earth revolved around the sun (Aristarchus), exploring the nature of diseases (Hippocrates), devising geometrical formulae and calculus (Euclid, Archimedes), and they established a taxonomy for plants and animals (Theophrastus). Schools of scientific thought, or natural philosophy, as it was called, were created by Socrates, Plato, and Aristotle. India developed schools of mathematics, astronomy, linguistics, medicine, and metallurgy (invention of stainless steel). China also founded schools of mathematics and astronomy, and one researcher developed a seismometer to enable people to prepare for earthquakes.

Classic Rome was formed as a republic. In 43 BCE, a triumvirate of Mark Antony, Octavian, and Marcus Lepidus became the leaders and divided the Roman territories: Antony governed the eastern territories, lands that had been conquered by Alexander the Great. Octavian presided in the Western provinces, including Italia and Gaul. Lepidus took over a small part of North Africa. War broke out among the three. Octavian was the eventual victor; he governed a new united Roman Empire. But a series of internal rebellions and uprisings, and then invasions by the Visigoths and the Vandals from the north of Europe led to the eventual fall of the Western Roman Empire.

After the fall, an intellectual decline took place in Europe. Fortunately, scientific activity had been taken up in the eastern Byzantine Empire and by Islamic scholars of the Caliphate in the middle east. In the meantime, Europe was becoming Christianized, and by the eleventh century most of western Europe was Roman Catholic. The Church became the home for re-emerging intellectual development.

The Renaissance, in the 15th and 16th Centuries, was sparked by Copernicus, Kepler, and Galileo, leading to the development of a heliocentric model of the solar system. Unfortunately, Galileo incurred the wrath of Pope Urban VIII by supporting the Copernican conclusions and refuting the "sacred" Aristotelian belief that the Earth was the center of the solar system. Later, Rene Descartes and Francis Bacon published philosophical arguments supporting non-Aristotelian science.

Led by Isaac Newton and Gottfried Leibniz, the Age of Enlightenment soon dawned, moving science into a more modern environment during the 17th and 18th Centuries. Scientific societies and academies were formed. Science became an important means for producing wealth, as well as inventions for improving human life. It also became widely popular among the people as the percentage of literacy increased. It included the famous dictum by Rene Descartes: Cogito ergo sum ("I think, therefore I am.")

Modernization of science accelerated in the 19th Century. Professional researchers began to replace amateur naturalists. Laws of the conservation of energy, momentum, and mass defined a highly stable universe. Darwin's 'On the Origin of Species' established evolution as an explanation of biological complexity. The electromagnetic theory by James Maxwell, defining the attraction between particles, took science beyond the laws formulated by Newton, and the discovery of X-rays by Wilhelm Roentgen led to the concept of radioactivity.

In the 20th Century, science became a major field of activity in terms of numbers of people at work, areas of research, types of institutions (public, private, and non-profit), and monetary investment. Remarkable discoveries were made. In 1953, the DNA molecule's double-helix structure was discovered by James Watson and Francis Crick, with contributions by Rosalind Franklin and Maurice Wilkins. Previously, it had been shown that DNA has three components: phosphate, sugar, and nitrogen containing bases. Watson and Crick showed how the components fit together: they built a three-dimensional model demonstrating how it might look. The molecule consists of two strands that wind around each other in a spiral. Each strand is like a backbone consisting of deoxyribose sugars attached to phosphates. Attached to the sugars are four nucleobases: adenine(A), cytosine(C), guanine(G), and thymine(T). Bonds between the bases form pairs, A with T and C with G, which hold the molecule together. All plants and animals, as well as some viruses, contain DNA molecules.

In mid to late 20th Century, the big-bang theory of the creation of the universe was formulated. The theory states that about 13.8 billion years ago there was no space, no stars, no planets, and no galaxies, only a void, in which a single, small, very heavy, and very hot floating particle existed. It suddenly exploded, creating the beginning of the universe. As the universe expanded, it began to cool, allowing the formation of sub-atomic particles and then atoms. These coalesced through gravity, eventually forming stars and galaxies. It was believed at first that the rate of expansion had been slowing, but cosmologists have found that the universe is expanding, at an increasing rate, which may be caused by a mysterious force called dark energy.

In 1957, the Soviet Union sent the first satellite into space. Eventually the USA also sent satellites up. Soon manned space flight became almost commonplace. A series of space stations were sent up starting in 1984. The International Space Station now in orbit went up in 1998. It is the largest modular station in low earth orbit, and is financially supported by USA, Russia, Japan, Europe Union, and Canada. Its mission is to pursue scientific research in astrobiology, astronomy, meteorology, physics, and other fields.

The phenomena of global warming and environmental pollution were shown to be serious dangers to the survival of the planet as we know it. These problems were described in Essays 3 and 4.

In the early 21st Century, cosmic revelations included gravitational waves and disturbances in the curvature of spacetime. In the molecular biology world, pluripotent stem cells with potential to develop into many different functional cells were revealed. The Human Genome Project, which determined nearly all the base pairs that make up human DNA, was completed.

These examples are just well-known samples of science research projects in progress all over the world, most of which function in relative obscurity. Nearly all may be expected to contribute information with some degree of significance.

To conclude this essay, it may be interesting and illustrative of the mysteries of science to consider dark matter and dark energy:

Dark matter is a hypothetical form of matter that may account for about 85% of the matter in the universe. Its existence is based primarily on gravitational effects not accounted for in the extant theories of gravity, especially because there seems to exist more total matter than can be seen.

Even more mysterious is dark energy. Cosmological observation indicates that the universe is expanding at an increasing rate, contrary to earlier concepts stemming from assessment of characteristics of the big bang, which suggest that the expansion has been slowing down from the beginning. Finding the opposite to be true shows that there must be an additional energy force that is unseen: dark energy. However, some experts in the field believe that a more accurate general relativistic treatment will show no need for an additional energy source.

7

Technology

One definition of technology is: the application of scientific knowledge for practical purposes. This definition has the advantage of simplicity but does little to describe the enormous scope of technological activities encompassed, or to its history beginning far back in time, predating the true human presence on earth. One could throw caution to the imaginary wind and include everything done by members of many animal species that involves some semblance of tool use. However, the technological cleverness of chimpanzees and birds are hardly relevant to or to blame for humanity's use of the various forms of technology for either benevolent or harmful purposes.

Still, we tend to think of technology as a series of developments resulting from scientific advances starting in the recent past and continuing in the present. However, I don't think that this definition gives sufficient credit to those who developed useful things during humanity's early years long ago with little or no help from previous ideas and inventions and certainly not from science as we define it. A list of some landmark inventions from the distant past might include the following: Stone points for cutting—about 1.7 million years ago; Use of fire—about 1 million years ago; *Homo sapiens* came along about 300,000 years ago; Artwork—100,000 years ago; Textiles—40,000 years ago; Shoes—40,000 years ago; Ceramic containers—20,000 years ago; Agriculture—10,000 years ago; Wine—8,000 years ago; Wheeled vehicles—5500 years ago; Chocolate—4000 years ago. In other words, many years ago, technological advances in a broad sense came before science.

Of course, as the distant past became the recent past, the population increased, as did, of course, the number of minds desiring to examine and solve problems. People learned from each other and from previous inventions, and the inventions occurred more often. Nowadays, on a worldwide basis, people are creating new inventions quite frequently, possibly every day.

The first major impetus enabling technology to generate enough forward motion towards its present rate of progress was the change from the hunter-gatherer life to the more sedentary occupation of agriculture, which in turn enabled urban growth and encouraged serious thought and planned activities. Let us examine the progress of technology during the development and fruition of the ancient civilizations of our history. The Mesopotamians opened the bronze age, working with copper, bronze, and iron (before the invention of carbon steel), which were used for weapons and armor. The Sumerians in Mesopotamia invented cuneiform writing. Cuneiform means wedge shaped, which describes the shape of the impressions made on a clay tablet.

In Africa, following the early development of tools by hominid ancestors, some of the earliest ironworking technology was developed, culminating in carbon steel production in high temperature blast furnaces in Tanzania, about 2000 years ago.

People developed and used simple machines: wheels and axles for potter's wheels and vehicles plus the level, pulley, and screw. The screw was used in early water pumps. A number system was developed, based on the number sixty (seconds in a minute, minutes in an hour, and 360 degrees in a circle). They had reached a relatively advanced level of mathematics.

The Egyptians invented the inclined plane for use as a ramp, which was vital for building their pyramids, along with wedges and levers. They developed maritime technologies for operating ships and lighthouses. Not least of their accomplishments was the use of papyrus for paper, as well as for reed boats, mats, rope, sandals, and baskets. Parchment, made from more flexible animal skins, eventually replaced papyrus for publications. Hieroglyphic symbol writing originated in Egypt as well. Writing was independently developed in China (about 1200 BCE) and Mesoamerica (Mexico and Guatemala, about 500 BCE).

True paper was invented in China, probably in the last two centuries BCE, and is credited to Cai Lun, an official of the Han Dynasty. After early use of hard materials to create a recording surface, a change was made. Hemp, flax, cotton, old rags, and ropes were among the original materials used in pulp form to make softer paper. Early in the history of the United States, the first paper mills used rag materials. Eventually, the primary materials were wood fibers from trees. A machine was invented that extracted the fiber from the wood, which was bleached after being pulped. Three inventions: the fountain pen, mass produced pencils, and the steam driven rotary press made the whole procedure easier and faster, enabling the introduction of cheaper paper usable for books, for use at schools, and for newspapers and magazines.

A series of industrial revolutions, the first coming in mid-18th Century, brought us to the information and cyber ages. We are now faced with a new age of worried wonder: What will the future bring?

The creation of a technological device is not quite the culmination of the inventive process. Rather it brings up some crucial questions: Can it be successfully used? Why and how will it be used, for good or bad, or both? Many inventions and discoveries have this dual life. Consider guns and gunpowder, radioactivity, money, knives, vehicles, social media, fossil fuels, pharmaceuticals, etc. I believe that the question is relevant for many technological advances. In addition, I believe it would be difficult to avoid the use of technology for negative use. There will always be people looking for nefarious ways to use new objects and ideas.

To repeat, technology has contributed to the development of our civilizations since before the word existed. We can possibly mark the rise of the desktop computer as the time when we began to see and understand technology's meaning, its history, and its importance as a force in our lives. In the early days of computers, many people had difficulty believing that we could progress past the room size computer, the ENIAC, to the desktop computer, let alone the laptop, smart phone, and smart watch computers we have now.

At the same time, it is well to remember the Luddites of 18th–19th Century England, workers in textiles and food production who rebelled against the mechanization of their work, which they felt would threaten their livelihood as well as increase the price of food. Ever since that time, worker v.

management conflicts have regularly broken out, and the appearance of new technologies has been the root cause of many of these conflicts.

In what manner has technology become harmful and a threat to society? One can speculate that the desire to do harm was latent among pre-humans and then humans before the existence of technology, when they needed only the personal instruments of harm to carry out their hidden desires. Their brains, arms, and legs were activated, finding ways to cause pain in others, for which, presumably they had some reason, such as envy, the need to chase away intruders, or to compete with other groups. Unfortunately, human brains learned how to look for opportunities to do harm, for one purpose or another. Eventually, governments assumed responsibility to arm many of its citizens, either to attack another country, or to be prepared to go to war if attacked. As time went on, industries sprang up to make various weapons of destruction to support the conflicts.

Technology became a major facet of society. In addition to war making, many technologies exist that contribute to global warming and pollution. We depend upon fossil fuels—petroleum, coal, and natural gas, the major producers of greenhouse gases, to power the engines that propel us over the land, water, and sky. Petroleum is also converted to other products—fertilizer, pesticides, and plastics, which are major sources of pollutants of soil, air, and water. The earth's few remaining massive forests are being converted to agriculture, ranching, mining, and logging, thus destroying valuable carbon sinks. We depend heavily upon animal products for our proteins and are rewarded with additional greenhouse gases, principally methane.

A third development in our world civilization, which we tout as the hallmark of technological progress, benefits large numbers of people but leaves many others out. That is the computer-smart phone-robotics-artificial intelligence world, which also creates economic, social, and opportunity inequality. Many people don't have skills that work with these forms of technology. Manual skills don't necessarily earn them enough money, status, and chance for advancement in this new world.

In the field of education, emphasis is being placed on the STEM subjects: science, technology, engineering. and mathematics, as the best means to learn skills valuable in the present marketplace. My own experience included attending an excellent, large, big city high school, which offered a wide variety of courses in English and foreign languages, history, economics, music, art, and civics, as well as mathematics and sciences plus a wide variety of extracurricular activities. My undergraduate and graduate school experience, culminating in a Ph.D. degree, encompassed what may be called a vocational education in genetics and plant breeding. I am pleased with the results, particularly since my experience also included liberal arts courses in literature, government, logic, economics, and the Spanish and German languages.

It is important, I believe, not to neglect the place of education to provide opportunities for those students, regardless of the reason, who do not have a STEM education. College graduates should have opportunities in the liberal arts, and those with a vocational education should have opportunities in blue collar careers that should minimize pay and benefit inequalities.

In the U.S. a broader need is the restoration of adequate public education funding by taxation, especially in the states, the principal sources of education support. Those funds were reduced after the 2008 recession in most states, and in several, very little funding was restored after the economic recovery. Teachers' salaries stagnated, schools failed to replace old and/or damaged books, and other necessities for an effective public educational experience were neglected—except in wealthier school districts.

Finally, we return to genetics and the bell-shaped curve describing the measure of intelligence in the population, using the Wechsler, Stanford-Binet, or other equivalent tests to measure general intelligence. General intelligence is a quantitative trait, distributed as a bell-shaped curve in a population. This type of curve typically shows a genetic effect modified by environmental influences. It demonstrates fundamental inequality in the population, which strongly influences essentially all the other inequalities that exist. (The bell-shaped curve is shown in the first essay.)

Advances in technology have occurred more and more rapidly in the years since the original industrial revolution in the late 19th Century. Each advance may open not just one new avenue of progress but several to many. Will we reach a time when the sheer volume of innovation will become overwhelming? Have we reached that point already? Certainly, large numbers of people reached that point years ago, and others will continue to succumb to the flood as the years go by.

The application of technology may have bad results in at least two ways. One is the result of deliberate attempts to cause harm. The other is the inadvertent consequence of some action. A famous example of the latter was the application of radium to the numbers on watch dials in an early attempt to enable a person to see the watch face in darkness. Many workers who applied the radioactive substance suffered from cancer after repeated exposure to the substance.

Technological advances for deliberate harmful use

were responsible for both the beginning and the end of World War II for the United States and Japan. On the morning of December 7, 1941, Japanese airplanes dropped bombs in a surprise attack on U.S. Navy ships anchored at Pearl Harbor Naval Base in Hawaii. They sank four ships and killed or wounded many American sailors. Four years later, American atomic bombs caused the utter destruction of two Japanese cities—Hiroshima and Nagasaki, with the loss of 129,000–226,000 lives.

More routine technology is our use of fossil fuels to power vehicles and other engines, warm or cool buildings, create electricity, and invent products such as plastics and fertilizers. The first people to use coal or a petroleum product to generate power almost certainly had no inkling of its effect on climate. This is no longer true.

Petroleum is one of the best examples of a product that is widely used despite its enormous capacity to generate greenhouse gases that are the primary causes of global warming and climate change. The oil companies knew, from their own research in the late 20th Century as well as undeniable results of research by climate scientists, that the evidence for global warming was real.

The CEOs and Directors who run the fossil fuel companies almost certainly must harbor many bad alleles of the applicable behavioral traits discussed in Essay One (greed, aggressiveness, extreme ambition, indifference, arrogance, dishonesty, and selfishness). Their scientists now produce results that conform to the company line. The companies financially support political figures who spout not only that global warming doesn't exist, but also accuse reputable and honorable climate scientists of perpetrating a massive hoax. The corporation managers and shareholders desire a great deal of money and wealth in the short term, and care very little about the future because they will be dead. They obviously don't care about the well-being of the next generations.

Humans who use technology to cause harm to other humans and to some animal species might consider the fact they are the nearly exclusive users of technology for that nefarious purpose. Most other animals use tools of various kinds to acquire food and build their nests They may also use one tool to create another. On the other hand, a troop of baboons will sometimes attack another to compete for food.

"Technological implements" have been found among several mammalian species: chimpanzees, bonobos, orangutans, gorillas, monkeys, elephants, mongooses, and badgers. On the surface of the oceans, and at varying depths, dolphins and sea otters are tool users. Some birds are accomplished users: finches, parrots, and especially corvids such as crows, rooks, and jays. Owls, gulls. herons, warblers, nuthatches, larks, and vultures also use implements, as do some species of insects. Most animals use tools for finding food, including capturing edible species, and if necessary, making the food more readily available for consumption.

Critics of technology (in the hands of humans) claim that technology itself will eventually bring catastrophe down on our heads, that it is a tool of capitalism. Others do not make that claim, stressing that the processes of industrialization can be pursued in a socialist society as well as in a capitalistic one. Once again, I claim the processes are based primarily on the distribution of genes and their alleles, defining people who are greedy, over-ambitious, selfish, uncaring, aggressive, and so on.

8

Belief Systems

It is fitting that a discussion of belief systems in this collection of essays should reside in juxtaposition to discussions of science and technology. They may touch on the same scientific phenomena, especially in the various fields of biology, leading often to heated exchanges, vocally and in writing, between proponents from each side.

There is a degree of complexity in some relationships. In the belief system grouping there may often be conflict between two or more related groups over beliefs or doctrines, each claiming to *know* the truth. This situation might occur commonly in discussions about religions, countries, political parties, races, and ethnic groups. Within the sciences, which are not belief systems, disagreements are usually over the design of experiments and interpretation of results.

A belief system can be described as a set of principles entrenched in the minds of people who share those principles, which are accepted as truths. Usually, the truths are not products of contemporary cognitive thought nor of experimental evidence. They may have been beliefs passed down from previous generations. The number of generations may vary from one or two up to many. Of the latter group, fear of black cats arose during the Middle Ages in Europe. Walking under a ladder as a bad luck action goes back to Ancient Egypt, about 5000 years ago. The 'truths' may also have been accepted by a person who simply wants to believe them, shrugging off the lack of evidence.

Why do belief systems arise? Probably the most defined belief systems are those of religion. The reason for that is the believer's deep conviction of the existence of an almighty deity, around whom has been built a structure of supportive beliefs. How did that concept arise? Judging from both pre-historic and historic evidence, one can imagine several stages. The first might be as an awareness of self and of others. This is a means of acknowledging relationships and similarity of beliefs, enabling members of a group to recognize that they had ideas in common, which may have bound them together, and at the same time maintaining their individualism.

The next stage might have been the death of an old member of a hunter-gatherer group, someone who had lived a seemingly very long time. He may have been expected to live forever. Yes, they thought, he seems to be dead, but perhaps not. Therefore, he must be somewhere out there observing us, perhaps even watching over us. (This idea was first expressed by historian Hendrik Willem van Loon.) It might have been the earliest concept of a personal god. Having achieved the idea of invisible beings, it would not be difficult to expand the concept to the presence of spirits, in the trees, the wind, and the waters, spirits who could control phenomena like the weather, the success of a crop of fruit or grain, the winning of a conflict, the behavior of animals, etc.

It would have seemed natural to give names to the various spirits, and to envision them with specific responsibilities. Ancient civilizations did just that. Starting about 5000 years ago, Sumer, Egypt, Greece, Rome, Scandinavia, Mesoamerica, and other ancient civilizations created pantheons of gods that the people could worship, and who could be called upon to aid in wars, protect crops, and even directly interact with the worshipers.

Ancient Egyptians worshipped a sun god named Ra, a special member of the pantheon of gods in their belief system. During the 18th Dynasty of the New Kingdom, the son of the deceased Pharoah Amenhotep III ascended the throne as Amenhotep IV in 1353 BCE. During the fifth year of his reign, he adopted a new belief that there was but one supreme god, Aten. Therefore, he changed his name to Akhenaten, decreed Aten's supremacy, and worshipped only Aten. At the same time, he downgraded all the other gods in the pantheon, especially Amun, whose cult had been the most powerful in Egypt. This did not please the worshipers of Amun and the other gods, and after Akhenaten's death, his son Tutankhamun returned Egypt to polytheism and attempted to rid the kingdom of references to Aten.

It appears that this religious adventure may have been the first one to glorify a single supreme deity. It was later echoed in Judaism, Christianity, and Islam, as well as Bahai and Zoroastrianism. Other existing belief systems have different sorts of thought and behavior rules.

Belief systems are not static. They vary with time, setting up rules that members of the group are expected to adhere to. Members of the group may become dissatisfied if the rules seem too harsh. They may rebel, which may result in their forming a subgroup, with new rules, or they may break away and form a new more moderate group.

The oldest Abrahamic faith is Judaism, which is the tenth largest faith group in the world, with 14–15 million adherents. It is based upon God's promise to Abraham to make a great nation from his offspring. God commanded the Jewish people to love one another and to worship only Him, and thus they would create a great nation. Over the years, the Jewish people divided into four main groups: Ultra-Orthodox, Orthodox, Conservative, and Reform. There are also non-believers, who consider themselves simply as ethnic Jews.

Christianity is the second of the Abrahamic religions, and the world's largest, with about 2.4 billion followers. The original Christian faith was Roman Catholicism, established at the behest of and based upon the teachings of Jesus Christ. After his crucifixion by the Romans, several converts, led by the apostle Paul, created a formal church in his name. Religious doctrines were written, some of which were later modified to refine the definition of the church. Other changes resulted in churches or rites that remained as Catholic entities, such as the Eastern Orthodox church. This separation occurred late in the 11th Century and is known as the Great Schism. Some disaffected groups broke away completely and are collectively known as Protestant churches. A complete description of the vast array of Christian churches might well require a very large tome and that is not my purpose in writing this chapter, which is a modest analysis of belief systems, in comparison to evidence-based systems, specifically the sciences. The Roman Catholic Church is the largest of the Christian churches, with 1.3 billion followers. Second are the collective Protestant churches with 900 million followers. Third is the Eastern Orthodox faith, with 220 million adherents.

The third Abrahamic religion is Islam, the world's

second largest religion, with 1.9 billion followers. Islam was founded in the Seventh Century by the prophet Muhammad, who taught that Allah is the one and only god. Muhammad wrote the Quran, which became the faith's sacred scriptures. After Muhammad's death in 632 CE, the Muslims split into two groups over choosing his successor. One group supported Muhammad's father-in-law, Abu Bakr and appointed him the first caliph as Muhammad's rightful successor. The members of this group are known as Sunni. It is the majority group. A second group believed that Muhammad had chosen his son-in-law Ali ibn Abi Talib as his successor. They are known as Shia and form a smaller group, which is centralized in Iran. Shia beliefs were influenced by Grecian ideas. For example, a Shia scientist named Ibn-al Haytham was a follower of Ptolemy, Euclid, and other Greek scientists. He was an outstanding mathematician and optics expert and is considered the originator of the scientific method and of the golden age of Islamic science.

Unfortunately, the golden age came to an end, probably as the result of conflicts among various political factions. It was a return to the belief system.

Returning to a more general discussion, several years ago I saw a pick-up truck with the following sticker in its rear window: *God did it, the bible says so, I believe it, end of argument.*

This struck me as a person's complete surrender to a belief system. It requires no evidence except that of a circular argument: God did it, the Bible says so, therefore God did it. There is, of course, a wide range in the depth of religious conviction, from absolute belief as illustrated above to casual belief of persons who may attend church occasionally, as a social event as much as a religious one. Many people move back and forth, from belief to non-belief and vice versa. Church leaders themselves vary in what they demand or expect from their parishioners.

One way or another, religions are primarily belief systems and distinguish themselves from other thought systems, especially the sciences, which need evidence to reach conclusions. The scope of required evidence to accept a conclusion differs considerably in the sciences compared to other non-belief systems, such as business organizations, governmental institutions, and so forth.

Two approaches for non-belief in God are atheism and agnosticism. The latter is often called a cop-out from atheism, but that is incorrect. An atheist believes that God doesn't exist. An agnostic doubts the existence of God but is not certain. For example, I am the product of a mixed marriage between an Episcopalian father and a Jewish mother. Neither was religious; they made no effort to steer me in any direction. As I grew older, I occasionally wondered if there was any reason to investigate joining the ranks of religious persons. At some point, I also began to ponder whether I should consider myself an atheist or an agnostic. The problem with atheism is that one must be as positive about the non-belief as of belief, also without evidence. In other words, atheism is a negative belief system.

The concept of a supreme deity is supported by two pillars: 1) the personal god who guides our lives, and 2) the creator of the universe. I find it impossible to believe that an all-controlling deity has been making innumerable decisions for humans and their pre-human ancestors, day after day, over millions of years. Judging from results, each of those supposed decisions might be described as wise, random, saucy, or cruel, and only the first one seems to be credible in an Almighty Being.

God as the Creator may be compared to a scientifically based creation concept. As a cosmological layman, I have tentatively accepted the Big Bang theory of the formation of the universe. This theory says that it all began with a small, very hot, very heavy particle floating in the void. Nothing else existed. The particle expanded about 13.8 billion years ago and the resulting super-particle went through a rapid expansion in a tiny fraction of a second, easily exceeding the speed of light. This may be explained by the void, which is a state of non-existence, and therefore had no resistance to the movement of the rapidly expanding universe. During the ensuing billions of years, a succession of products of the event were formed: subatomic particles were first, followed by atoms. These eventually consolidated into larger entities, forming proto stars and planets, and eventually became the entire universe. This sequence of events requires us also to accept that the original small particle might have been there forever. Words like forever and infinity, unfortunately, are undefinable. They don't really have any meaning that can be grasped. On the other hand, if a deity was the creator, where did the deity come from? Believers say that the creator has existed forever. There is that word again. I don't have any suitable answers to the question, and I doubt that anyone else does. Therefore, I am an agnostic, meaning I do not know the answers.

Returning to the religions of the world, the Abrahamic faiths are most common in the Western continents, although there are many people of these faiths in other parts of the world as well.

Several religions have predominated in the eastern and southeastern countries of Asia. Hinduism is the principal faith in India and the third largest in the world, with 1.25 billion followers. It serves both as a faith and as a dharma, which means "right way of living". There are also significant numbers of followers in Mauritius and Nepal, as well as countries of the Americas, Africa, and others.

Buddhism is the second largest religion in India, and the fourth largest worldwide. It is also important in other countries of Southeast Asia and East Asia. It was founded by the religious teacher Gautama Buddha and is based upon spiritual practices to end suffering. Taoism is a system that teaches about various disciplines for achieving perfection through self-cultivation. Taoism and Buddhism are important in China, as well as Confucianism, named after the ancient Chinese philosopher, Confucius. Under the present regime, it is not known how many people in China practice those faiths. Shinto is a polytheistic faith practiced principally in Japan.

The most recently founded major faith is Sikhism, which appeared on the Indian sub-continent near the end of the 15th Century. It is based on the spiritual teachings of Guru Nanak. The core beliefs are faith and meditation in the name of the creator, equality of all mankind, selfless service, justice for all, and honest conduct. It is the fifth largest religion worldwide.

Some existing religions are polytheistic, as were most ancient ones. Some modern belief systems are still based on the presence of spirits, in humans, animals, plants, and even non-living things.

In addition to religion, the most notable of belief systems, other social groups also subscribe to belief systems, either deliberately or by subconsciously developing standards to which people may adhere. These include political parties, ethnic groups, racial groups, gender groups, nationalities, social standing societies, and other real or imagined groupings. Some of these groups have almost as much emotional drawing power as religion. Racial groups have relationships that are clouded by senses of superiority and inferiority by which one group looks down upon another, creating feelings of animosity.

For example, beyond the comfort derived from kinship with others, members of these groups often tend to develop attitudes derived from their willingness to draw conclusions with minimal evidence, which too often lead them to denigrate people of other groups. Such attitudes have repeatedly descended to dislike, hatred, and violence.

Development of a human belief system often derives from simple ignorance. Many thousands of years ago, people did not know much about anything. Each generation learned a little bit more about life, death, and the world around them, until civilizations blossomed that were based largely upon the advent of agriculture. The rate of learning increased ever more rapidly for the next ten thousand years or so. Here we are now, stuffed with knowledge. Unfortunately, the rate of learning, or even the willingness to learn, varies among and within groups. Many people became content to rely on past knowledge, right or wrong, because it was easy and probably felt good. Enter the belief systems. Some, like those mentioned in the previous paragraphs, are large in scope, and strongly influence many people. Others are minor and send people to astrologers, force builders to create hotels lacking a thirteenth floor, or induce people to dodge black cats and avoid walking under ladders.

The theme of this book, as discussed in the first chapter, is the effect of genetics and evolution on the behavior of humans: How do we treat the planet, the various non-human creatures that reside on it, and our fellow humans? Some of the widespread strongly felt belief systems have often led us to warfare and other types of violence because of racism, gender discrimination, religious and ethnic persecution, political animosity, and other extreme prejudices. Strongly held emotionally charged beliefs too often are unsupported by evidence and are accompanied by a lack of interest in hearing the other side of an issue.

Religion, nationalism, ethnicity, and race are probably the belief systems that most easily arouse members of one group to commit hateful and violent acts against another. Groups that are less formally organized are less likely to generate the same deeply held feelings as those four systems. Sports fans, college alumni, and similar groups may generate temporary emotions against rivals, but few people in such groups are likely to remain in a permanent state of anger or hatred caused by the social strains.

The possible consequences stemming from religious, nationalistic, ethnic, and racial malevolence and bitterness include persecution, deprivation of rights, exclusion, personal injurious attacks, and mockery, as well as various types of conflict: wars among nations over territorial supremacy, civil wars, so-called ethnic cleansing attacks, and religious conflicts.

Belief systems arise when people fail to adopt new ideas or are deprived of the opportunity to acquire knowledge that has been accumulating throughout the millennia of human existence. Instead, they may rely on untested "truths" from parents, friends, holy books, leaders, teachers, propagandists, and others who obtain information from various sources that are claimed as truths, often without any evidence. The bumper sticker quoted above captures the gist of the claims very nicely.

As this paragraph was first being written, Donald J. Trump was President of the United States. He harbors a truly remarkable array of belief systems that are either shared with others or seem to be uniquely his own. I had no difficulty labeling him with thirteen of the detrimental traits of behavior listed earlier:

Greed—Maintaining income from his properties during his presidency.

Arrogance—Stating that he can do anything he wants because of his position. He claimed he could shoot someone on Fifth Avenue in New York City and get away with it. He sometimes claims he is the only one who can solve existing problems.

Aggressiveness—Forcing his way to the front of group photographs.

Intolerance—Disdain for people of color. Suggested that we should only admit immigrants from Norway and the like: good white people.

Credulity—Considers global warming a hoax despite an abundance of evidence to the contrary.

Irrationality—Believes himself to be a "stable genius." Told Hillary Clinton in a 2016 debate that he would send her to jail.

Violence—Publicly brags about grabbing women by their private parts.

Cruelty—Authorized separating immigrant children from their parents, while being held in border camps.

Dishonesty—Fact checkers in the press have identified thousands of his untruths.

Ethics—Appropriated money donated to his charity for his personal use.

Wariness—Believes that liberals and the press are out to get him.

Jealousy—Denied President Obama's accomplishments and wanted to eliminate Obamacare.

Extreme ambition—Built an enormous real estate

empire. Praises powerful autocrats and probably wants to become one.

He faced the U.S. Senate on trial after being impeached by the House of Representatives for abuse of power and obstruction of Congress. A whistleblower declared that Trump tried to persuade the President of Ukraine to investigate Joe Biden, Trump's Democratic rival for the presidency in the 2020 election, and his son Hunter for an alleged conspiracy to influence the 2016 election. Trump threatened to withhold military aid otherwise. The purpose was to switch the blame for the 2016 election interference from Russia to Ukraine. There was no evidence to support his allegations. The issue went to the Republican controlled Senate. To convict him would have required a two-thirds majority, which did not happen.

I think Donald Trump was a very bad president and his election was a definite triumph of nonsensical political belief systems. It would be a blessing if he disappeared from the political world. Unfortunately, he is running for president in 2024, and at this writing he is the favorite to be nominated by the Republican Party.

However, in 2023 he faced four criminal indictments: New York: 34 criminal charges of falsifying business documents related to payments made to Stormy Daniels before the 2016 presidential election; Florida: a federal criminal indictment regarding mishandling of sensitive documents and conspiracy to obstruct the government in retrieving these documents; Washington, D.C.: a federal indictment of four criminal charges of attempting to overturn the 2020 election of Joe Biden; Georgia, an indictment of 13 criminal charges to overthrow Biden's victory in Georgia, with 18 accused co-conspirators.

One belief system is unique in its single-issue

foundation. The Second Amendment of the U.S. constitution states the following:

"A well-regulated Militia, being necessary to the security of a free State, the right of the people to keep and bear Arms, shall not be infringed."

I find it difficult to find a rational basis for the National Rifle Association and others to ignore the first thirteen words of that sentence. Of course, there is no rational basis. They simply ignore it because they want to. It is a single-issue belief system, albeit an unusually powerful one. At the time the amendment was written, it was feared that the states might be attacked by a federal army and therefore would need the protection of individual state militias. This is no longer a meaningful expectation. None of our states seriously worry about being invaded by the U.S. Army, although there are a few people who do believe in that possibility. There may be legitimate reasons to possess a firearm that don't unnecessarily lean on the Constitution: limited hunting, defense in dangerous areas, target practice in locations created for the purpose.

Unfortunately, the Trump traits described above seem to be intent on bringing violence front and center in dealing with the objects of his scorn, aided by sections of the Republican Party who have been purging the party of its moderate members for many years. The tentative predictions I am making in this book may come true sooner than later. Consider the January 6, 2021, insurrection by ultra-rightwing groups. A large mob invaded the U. S. Capital Building, leading to five deaths and many injuries. Trump probably could have stopped the insurrection, but he chose not to because he organized it.

Donald Trump is a natural outgrowth of the grotesque changes in the Republican Party that have occurred since Lewis Powell, a lawyer, and future Supreme Court Justice, wrote a letter to the U.S. Chamber of Commerce admonishing business organizations to become more active in politics to help attain the same level of influence as liberal organizations, but to speak out for business interests. It appears that the response went far beyond that recommendation, leading to the Citizens United decision of the Supreme Court, enabling the very rich to make undisclosed contributions in the high millions of dollars to political causes. The result was that a small minority of people, the very rich, too often control the outcome of elections and insure the faithful loyalty of office holders, from low level political positions to high offices. Unchecked, this will lead to the loss of our democratic institutions. I recommend a book by Jane Mayer, "Dark Money" for an excellent portrayal of the billionaires who donate millions of dollars to achieve the elimination of government for being in the way of these greedy people.

In early days, there were only primitive belief systems until some people learned to rely on evidence without realizing that they were adopting a scientific point of view. In time, the study of nature became known as natural philosophy as espoused in ancient Greece by Socrates, Plato, Aristotle, and others. This designation remained, covering many systems of the natural world, until the 19th Century, when the term science was adopted. Thereupon, natural philosophy became science and technology, with conclusions drawn when supported by evidence.

9

Government

There are nearly two hundred countries in the world and as many governments. Therein lies one source of many of the world's troubles, among and within countries. A related source lies within us. That source is the tremendous variation in genetic make-up among the world's eight billion people. As explained in Essay One, no two persons are genetically the same, except identical twins. Both sources contribute to great variability in outlook of human affairs, which often leads to conflict, stemming from different individual viewpoints on national policies.

Governments have many responsibilities. They control, mediate, or referee many different issues: tax bases, regulations, support for organizations, business affairs, educational institutions, infrastructure, administration of parks and recreational areas, policing, transportation, constitutionality, and foreign affairs. Foreign relations range from friendly contacts to conflicts, including warfare.

The countries of the world are different from each

other because of variation in their histories, their geographies, their resources, their population densities, their belief systems, and their evolutionary paths. Consequently, they consist of democracies, dictatorships, oligarchies, and near anarchies. Within each group, there may be republics, monarchies, and theocracies, which are overlaid by capitalistic or socialistic economic systems. The number of combinations of these characteristics is large, and within each group are differences of degree. Within each country, there is, of course, much genetic variation.

Governments of the world's countries vary, influenced by the above characteristics, and range from terrible to excellent. Between those extremes are those that are not so terrible, slightly terrible, middling, pretty good, good, and better. A key characteristic, of course, would be the degree of power hunger in the leadership. Governments vary in a similar manner as people, based on their genetic variation. Therefore, we may ask the question: How do governments contribute threats to world survival? What sort of actions do they take that would lead us down that path to disaster? One obvious action is to start a war. In the past, no war was fought with a powerful enough cumulative force to lead to world disaster, with at least two possible exceptions.

World War II could have concluded in planetary disaster. It did not happen, as it turned out. However, it is well to consider that a distinct possibility hovered over the planet that could have had more dire results than it did. Nazi Germany had a group of scientists working to create an atomic weapon. If Germany had developed the weapon before the Allies, it is not beyond reason that it would have been used, creating unbelievable destruction, fear, and horror in the world.

What would have been our response? The Allies

could have surrendered; this might have been followed by the fulfillment of Hitler's supposed goal, "Tomorrow the world!" Or perhaps they might have been close enough to completing a bomb to attempt retaliation. Result? A nuclear war perhaps, although it might have been of limited scope because the number of nuclear weapons that existed at that time probably would have been insufficient to carry out a lengthy war. However, it might have led to a world in chaos. There is little comfort to be imagined in any subsequent scenario.

Nonetheless, we did build nuclear weapons first and used two of them in Japan to end World War II. Japan had no comparable weapon or other means to continue fighting. The cost in lives and destruction was beyond previous conception. Was our reasoning for dropping those bombs acceptable? The debate over that question continues to this day.

A second nuclear challenge was the Cuban Missile Crisis in the fall of 1962, which brought the United States and the Soviet Union to the edge of a nuclear conflict that could have wreaked unbelievable havoc on the world. In the years since the end of World War II, both countries had built considerably bigger stockpiles than had previously existed. After discovering that the Soviet Union had accepted Cuba's request for missile sites on the island, President Kennedy ordered a naval quarantine of Cuba rather than an air strike followed by an invasion. After several days of tense negotiations between Kennedy and First Secretary Nikita Khrushchev, it was agreed that the Soviet missile sites would be dismantled, and that the United States would dismantle Jupiter missiles in Turkey. The key move for Kennedy was to accept a mild conciliatory message rather than responding to a threatening one from Khrushchev. The whole confrontation took place in an atmosphere of escalating tensions between the two countries.

Possible future scenarios are more frightening. Nine nations now have nuclear weapons, with stockpiles ranging up to several thousand warheads, totaling about 15,000. No bombs have been dropped as weapons since 1945. Since then, the world has existed under a bizarre arrangement called MAD: Mutually Assured Destruction, which assumes that each nation with nuclear weapons will refrain from using them for fear of massive retaliation. We have lived under this sword for seventy-eight years at this writing, with no guarantee that it will not fall on our heads at some future time. A new world war may create situations in which nations fashion a mindset that overcomes doubt, and one country will declare "First strike now!" With that decision, all will probably be lost. It cannot be allowed to happen.

Governments are also basically responsible for fueling runaway climate change. As noted earlier, President Trump withdrew the United States from the Paris Climate Agreement, which supposedly limits the global temperature rise to 1.5 or 2.0 degrees Celsius. The withdrawal took effect in November of 2020, the day before the Presidential election, which Trump lost. New President Joe Biden restored the US membership when he took office in 2021.

Overall, CO_2 emissions have continued to rise but hopefully at a slower rate in recent years. However, Jair Bolsonaro, former president of Brazil, gave his blessing to destroy much of the Brazilian rain forest to accommodate farming, ranching, and mining interests. That forest is the largest CO_2 sink in the world; its loss would bring the world closer to the catastrophic scenario referred to as "Hot Earth".

One of the problems with the agreement protocols is that goals and methods are determined and set, not by scientists who fully understand the danger, but by each nation's political leadership, whose decisions are often dependent upon the preferences of internal commercial interests. A consensus of climate experts predicts that the average global temperature rise will exceed the target of 2^0 Celsius above pre-industrialization by as much as $2–3^0$ C. by the year 2100, which will insure a "Hot Earth" scenario. Many islands will be submerged, and coastlines of every continent on the planet will recede. A substantial number of major coastal cities and towns in the world will go partly underwater. It is imperative for us to exert greater effort on a world scale to avoid that catastrophe.

To repeat, it is advisable, and frustrating, to remember that the guidelines on this issue were developed by climate scientists whose mission it is to set them in the best interests of the planet and its inhabitants. Unfortunately, implementation of the guidelines has become a political issue, interpreted by governments in different ways, including commercial interests as well as the scientific findings. This essentially bypasses the guidelines, of course.

Governments play roles, of varying degrees of influence, on the contribution of science, technology, education, business, and agriculture as practiced in their countries. A government's major role is that of monetary support, based upon the imposition of taxes, tariffs, and fees on income, sales, imports, land usage, inheritance, corporate profits, and investment profits. At present, the level of support for the sciences, technological institutions, schools and colleges, industries, and farmers varies depending upon the perceived needs and desires of the political parties in power. Liberal governments lean more towards support for the conclusions and advice of science and technology, while conservative governments prefer to support business. The level and type of support, therefore, may vary considerably among countries.

In addition to climate science, a liberal-conservative division exists regarding government regulation of other practices, particularly by commercial establishments. Businesses prefer little or no regulation by the government; they would rather self-regulate, which often means not at all. This is not surprising, because regulation means extra costs, which decrease company profits. I would call this a commercial behavioral trait: indifference to needed standards and regulations.

Another responsibility that includes the participation of governments is the setting of national policies that determine whether equality or inequality, in a broad sense, is the hallmark of their societies. A government consists of a group of departments, or ministries, to carry out the executive missions. The heads of these departments are expected to believe in the mission of their departments. Sometimes, in politically conservative governments, the leader deliberately appoints people who are against a department's mission and will endeavor to reduce the regulative activity of the department, thus pleasing the leader and his commercial supporters. Refer to the Trump administration for evidence of this sort of leadership, as well as similar conservative leadership in other countries.

Governments may become aligned with several different belief systems or national institutions, deliberately or unwittingly, which might lead to unfortunate results. Religion is the foremost belief system of this sort. In ancient times, Egyptian pharaohs, Sumerian kings, Greek and Roman emperors, and others were closely attached to their gods. In those countries that later adopted Roman Catholicism, governments essentially merged with the church to share political power to the point where the Popes were sometimes the dominant forces. Similar relationships developed in some Islamic nations, where caliphs once ruled as successors to the prophet Muhammad. In the Soviet Union, China, and the smaller Communist countries, the Communist Party had the same function as the Catholic Church and Islam: partner with the governments. This type of partnership has also occurred in the Israeli government.

Protestant churches often advocated for religious freedom and separation of church and state, which carried over to the settlements in the Western Hemisphere and elsewhere. However, evangelical movements rose, starting in the 18th Century and continuing until the present time. Aside from preaching fundamentalism in the interpretation of the Bible, some evangelical churches became allied to governments as virtual state churches. Evangelicals, in the United States and elsewhere, lean rightward politically, to the Republican Party in the U.S. and similar parties elsewhere.

Government can be viewed as evil incarnate or as a savior, or descriptions between those extremes. Assessing the governments of the world, historically and at present, one can indeed see the variation and recognize that the peoples of the world have existed historically under one kind or another. Our acquaintance with genetics explains many of the differences and is the possible basis for doing a better job someday.

At the present time, the elected governments of the world are partly products of the collective genotypes of their leaders and partly of the genotypes of the citizens of their countries. The more autocratic the leadership, the fewer contributions come from its citizens. If the leaders had come from the ranks of people with benevolent tendencies, due to favorable genotypes, then our histories, our present outlooks, and our future expectations would be far different. Too often, leaders come from the ranks of those who would be warriors and conquerors, and they are often cruel, greedy, selfish, overambitious, uncaring, arrogant, and so on.

There is one type of government that does not exist now and has never existed: a world federal government. This enormous change may come to pass in the future, and if so, would be an important key to our survival. I will discuss this in the final essay.

At present, the governments of the world can be divided into several categories by political type. These include presidential republic, semi-presidential republic (shares leadership with a prime minister), republic with executive president elected by the legislature, parliamentary republic with a ceremonial president, constitutional monarchy, constitutional parliamentary monarchy, absolute monarchy, one party state, country with suspended constitutional provisions, country with a transitional government, and country without its own government.

The previous paragraph suggests one of the difficulties in forming a world government of any type. That difficulty is the large number of substantially different types of existing governments: democratic, communistic, fascist, royal, tyrannical, etc. The need to choose, preferably, a democratic type of world government by representatives of many other types will be a formidable obstacle. It might be less formidable if the world was in such bad shape that the leaders finally saw that they had to put aside their differences long enough to take the enormous step into a new harbor of salvation. It would be even better if some governments would see the light, so to speak, and become more democratic after recognizing that to do so would lead to a greater chance of survival in a better world.

After World War I, an international organization was formed at the initiative of the victorious allied powers, especially U.S. President Woodrow Wilson. It was called the League of Nations and was the largest world-wide organization of its kind with the mission to preserve peace. Its principal goals were to prevent war through collective security and disarmament by settling international disputes through negotiation and arbitration. Although Wilson won the Nobel Peace Prize for his contribution, he was unable to persuade the conservative Republican Party to support America's membership, and the USA did not join the League. The League was unable to deal consistently with the international problems, including preventing aggression by the Axis powers of Germany and Italy, which, along with Japan, quit the League. These and other difficulties led to World War II, starting in September 1939. When that war ended, a new international organization, the United Nations, was formed. Its overall success is again uncertain.

10

Communication

It may be difficult to imagine communication and education as threats to survival, but a little thought plus examination of the idea might well show the way. Both are means of transmitting ideas and information from one group of people to another. The transmitting group is usually smaller than the receiving group, which enables it to multiply its impact, thus becoming an instrument of substantial influence. Also, the transmitting group possesses the appropriate words and ideas, which add to its authority. Therefore, educators and other communicators are the principal holders and transmitters of ideas and information in their various fields.

At some unknown time in the ancestry of humans, two or more individuals discovered that they could communicate with each other using sounds or pictures, or both, thus making their activities easier and more effective. The activities included acquiring food and water, finding shelter, hiding, fighting, and/or traveling. Now, many thousands of years later, communication is an encompassing word, a rather enormous tent that covers many means of transmitting information: from conversation and lectures to radio, television programs, and movies to advertising, to books, and so forth. The manner of communication may be expressed through facial expression, whispers, hints, talking, and shouting appropriate to the medium of the communication.

How does communication fit into the theme of this book as a danger to our survival? Does it not enlighten our thoughts and ideas, thus offering us a means to avoid misunderstanding and conflict? It does if those are the goals. But if they are not, if the idea is to accuse, to foment hatred, to destroy a reputation, to mislead, to plan a genocide, and on and on, communication offers many outlets for people with a set of bad behavioral alleles. Communication can augment the force of the other issues in this book. It can do good or great harm. It offers many tools useful for a few people or for many.

From the beginning, communication offered a method to transmit ideas, directions, explanations, questions, and answers, as well as feelings of joy, sadness, anger, or pain. Little has changed in those basic opportunities. What has changed are the additional methods beyond the basic ones, methods based on ever advancing technology, offering more and more highly sophisticated tools. The impetus has been the burgeoning population throughout our history, which eventually demanded means of mass communication reaching populations all over the world.

However, one purpose of this essay is to show how the various forms of communication may be used to threaten world survival. Here is an example from the not-too-distant past. Adolf Hitler posed one of the most dangerous communication threats in history. Born in Austria, he migrated to Germany, a nation he worshiped, at a young age. After service in the German army during World War I, he enhanced his hatreds, against Jews, Communists, and other groups and populations. He was arrested in 1923 for inflammatory speeches. While in jail, he dictated the first volume of Mein Kampf, the statement of his beliefs, to Rudolf Hess, a Nazi Party comrade. After several years spent strengthening the Party, Hitler became active in politics and was eventually appointed as Chancellor and then Fuhrer.

In 1939, he deliberately plunged the planet into World War II, which soon became global and resulted in the largest number of war fatalities ever. The war ended after six years, with the destruction of two Japanese cities by the most powerful weapon ever devised at that time: the atomic bomb. As noted earlier, Germany had attempted to build a nuclear weapon but had not been able to complete the construction before being overrun and forced to surrender unconditionally early in 1945.

Hitler, with Minister of Propaganda Joseph Goebbels and family, had taken refuge in the Fuhrer's underground bunker in Berlin. As the Soviet army advanced into the city, the refugees realized that the war was lost. Hitler shot himself. His new wife Eva Braun took cyanide. Goebbels and his wife Magda took cyanide and shot themselves, after they murdered their six children. So ended the lives of the Fuhrer and his Minister of Propaganda after only twelve years of the projected thousand year Third Reich.

To further the Nazi cause during those twelve years, Goebbels exploited for propaganda purposes all major means of communication current at that time: films, radio, newspapers, books, art, music, and theater. He also undertook censorship and book burning. The propaganda was meant to glorify Hitler, discourage people from independent thought, denigrate 'degenerate' art and music, foster anti-Semitism, and present the world with an image of a united German population. The propaganda campaign was also successful in some countries outside of Germany, generating support by small groups for the Nazi regime. Hitler was admired by some for bringing law and order to his adopted country, and many shared his abhorrent anti-Semitism. The groups entranced by Nazi Germany included people in the United States and Great Britain, soon to be allies that fought against the Nazi regime.

Various forms of neo-Nazism have recently shown up in Europe, the United States, and other parts of the world, beginning shortly after World War II. It has continued to the present day. Is this a carryover of Goebbels' sickening propaganda? I suspect that it is, although long time prejudices in each country continue to play their part. For example, the Ku Klux Klan, an American fascist organization, has had two incarnations since the end of the Civil War and the reconstruction period. Its primary message is directed against African Americans, but anti-Semitism, anti-Catholicism, and anti-foreign-born-minorities are also part of its hate list. The Klan has used lynching, cross burning, and other terrorizing threats, primarily against African Americans.

In autocracies, news and comment outlets on television, radio, newspapers, and magazines reflect the political and social beliefs of the ruling party, usually in extreme tones. In democracies we expect viewpoints from both liberal and conservative commentators, and we hope that both sides present reasonable points of view. Unfortunately, in the United States, the Fox Network and other similar television outlets and radio talk-shows present ultra-rightwing propaganda, composed of hateful lies and distortions far beyond conservative points of view.

Most disturbing in the current American political scene is the descent of the Republican Party. The party seems to have lost interest in governing, preferring instead to run for offices at various political levels, working to prevent people of color from voting, and encouraging violence in support of lies and distortions, including Donald Trump's denial that he lost the 2020 election and his provoking of the January 6th mob insurrection at the US Capitol Building.

Education properly used is a premier form of communication. Education encompasses all the types of communication, and more when we consider not only the specific tools of education, but also many informal methods. We educate ourselves and others by reading, thinking, talking to other persons, talking to ourselves, and wondering about things.

Obviously, education began in prehistoric times, as a means for adults to pass on skills to each other and to the young, enabling their survival and well-being. Over time, as cultures developed, more formal methods were employed. Schools were founded during the Middle Kingdom period of ancient Egypt, over four thousand years ago. Plato created his Academy in 387 BCE in Athens. It was the first institution of higher learning in Europe. Educational institutions continued to develop in Europe until the fall of Rome in the late Fifth Century and the beginning of the Dark Ages. The Catholic Church then became the only home for literate scholarship in Europe until the Renaissance. During that period, formal education persisted in China, based on the teachings of Confucius. In the Middle East, the Islamic Caliphate continued to support education, and in the New World, the Aztec culture of the late period of pre-Columbian years had a well-developed system of learning.

Over the years, formal education became entrenched in most countries of the world, usually at several sequential levels: primary school, secondary school, college, and graduate school. School teachers provide information about various subjects to children and young adults, who are usually relatively naïve about the extent and availability of the transmitted information. Other educational sources include books, magazines, newspapers, radio and television broadcasts, public speakers, and in more recent times, tweeters, who pass along not only information, but also their interpretations of its meaning. The practice of tweeting has achieved a reputation of passing along nasty comments that often go viral.

In democracies, receivers of information and its meaning from teachers and communicators expect that they are learning the truth, that their education is fulfilling a commitment to acquaint them with not only the three Rs, but also an unbiased view of history, languages, sciences, and mathematics, as well as a view of the world around them, and other subjects to help them choose their places in the world.

In non-democratic societies, students may have similar expectations, but are more likely to receive biased information masquerading as truth, particularly in classes dealing with religion, history, world affairs, or economics, which are designed to favor the single political party running the government. Also, even in democracies, the information brought to students, as well as listeners and viewers of the various outlets, may be colored by the teacher's previously acquired information in his/her own education and from beliefs acquired from the media. Subtle or drastic variation in the presentations by communication outlets may be confusing to readers, listeners, and viewers of the media. Differences in content or mode of presentation by teachers may be confusing to students, or they may be quite useful by enabling them to weigh more than one source of evidence and come to their own conclusions. This opportunity is likely to be absent in an autocratic political climate.

Regardless of the subtleties, on an international level the differences noted above may contribute in a basic way to the building of bad feeling among the various countries of the world about each other, possibly leading to anger and disagreement over the conflicting views, and then to armed conflict. A violent conflict between two countries could be extremely dangerous if they both possess nuclear weapons. Also, other countries that may possess nuclear weapons could be drawn into the conflict. In other words, once the guns are fired, and the highly dangerous weapons join the conflict, the threat to survival raises its ugly head.

Education and communication differences alone are not likely to lead to catastrophic conflict. But those phenomena do not exist in a vacuum, as illustrated in the other essays of this volume. It is especially important to maintain awareness of the basic underpinnings of genetics and evolution that generate tremendous variation in understanding and interpretation among the billions of humans residing on the planet.

During the first half of the 20th Century, the principal methods of communication to the public were radio, newspapers, magazines, and news shorts in movie houses. By the middle of the century, television joined the group. Together, they covered the news, including national, international, and local events in various venues. Paper outlets also presented features such as puzzles, gossip, and comic strips. These activities were supported by subscription charges, purchases from newspaper stands, and advertising. Many communities, especially the larger cities, had multiple newspapers and TV and radio stations.

Late in the century and early in the 21st Century additional outlets were added: Email, streaming, smart phones, smart watches, and social media. These information transmitters allowed instant communication, which was good in the sense that non-professional people and institutions could also transmit news, ideas, questions, and advice. In addition, propaganda could be transmitted, either with good will, or with negative purposes, communicating mean spirited lies and accusations, a process called trolling, designed to hurt enemies, thus becoming a political weapon that is difficult to refute effectively.

The modern terms and devices used in communication have had major effects on the political structure and on methods of persuasion. Modern methods have both advantages and drawbacks. For example, the various social media outlets have the advantages of speed and very wide coverage. Messages can go viral; they may reach thousands, perhaps millions, of people in a short time. Printed news, television, radio, and mail cannot do this. Such rapid coverage is wonderful when the message is a benevolent one, at least for those who agree with the premises of the message. If the message is wicked, mean spirited, and/or harmful in some way, it is not wonderful.

An unfortunate consequence of the rise of the newer forms of communication and the lesser use of the traditional ones is that many people have been cancelling their subscriptions to the slower newspapers and magazines. Publishers have reduced their staffs, cut salaries, narrowed their focus, and reduced the size and coverage of their publications. Many have gone out of business. Many readers lament the reduced number of news stories, and the loss of features such as local and national columnists. Fortunately, much of the material is available electronically for people possessing the right equipment.

Additionally, there has been a mushrooming of ultra-right-wing radio and television outlets devoted to smearing liberal persons, organizations, and ideas. They have put to the test the old saying that a lie, repeated enough times, becomes the truth, at least to those who are already receptive. Unfortunately, this has happened with the followers of the Fox Network and other sources on the far right. There are fewer left, liberal outlets to compete with the right, largely because the owners of commercial outlets are business owners, most likely to be right wing people themselves.

Communication is a valuable but potentially dangerous tool that can augment the threats described in other essays in this book. The danger had been illustrated too well by the Trump administration during the unfortunate four years of his disastrous presidency.

Hopefully, four years of Trump will be all. At this writing (May 2024), he is running for a second term and is the favorite among several Republican candidates, although he is under multiple civil and criminal indictments, the most serious of which are his criminal attempt to change the results of the 2020 election, which he lost to incumbent president Joe Biden, and his part in causing the January 6 insurrection.

11

Business and Commerce

Business is an exchange activity that exists to sell goods and services to people, businesses, governments, and other institutions, with the primary goal of making a profit and sometimes of providing various social benefits. Commerce is a subset of business, pertaining to the distribution of goods and services.

Businesses vary tremendously in size and scope. On the small side are one or two person "mom and pop" stores ranging up to those with a total of fifty employees. Companies with 51–250 employees are considered medium in size. Businesses with more than 250 employees are large to very large. However, the number of employees is only part of the description. Perhaps more important is the level of income. If one looks at a list on the internet of the fifty largest companies in the world that have published their financial data, all of them have gross incomes of many billions of dollars, while the number of employees varies from about 4000 to over 2 million. Most are privately owned companies, but some are state owned. Many of them are global in scope.

Medium and large businesses are structured with high level management, mid-level management, and worker positions. The income differential for the three groups varies from very high down to very low earnings. This differential describes economic inequality on a stark scale. In addition, large corporations are partially supported by the sale of company stock to outside shareholders, many of whom own thousands of shares and receive dividends worth large amounts of money. This also contributes to inequality.

Economic inequality is one type of inequality that contributes to unrest among disadvantaged populations, which in turn may lead to active conflict, within or between countries.

One of the very important characteristics of the business world has been the extensive growth and development to a globalized level, from non-existence many years ago to its present world scope. The growth may have started among the early civilizations of the Middle East, in Sumer and the Indus Valley, during the Third Millennium BCE. Additionally, in the 2nd Century BCE, a network of overland routes called the Silk Road became the major connection between East and West, bringing together China, India, Persia, southeastern Europe, and Arabia. It functioned not only as a means of transporting hard goods and people, but also spread cultural attributes, including religion, art, languages, new technologies, and political ideas, especially with the massive expansion of the Mongol Empire. Led by Genghis Khan and his sons and grandsons, the Mongol Horde built the largest contiguous empire ever, totaling over 23 million square kilometers, and covering much of Asia, Russia, the Middle East, and southeastern Europe during the 13th and 14th Centuries.

From the 16th to the 19th Centuries, globalization took the form of expansion of the maritime empires of Spain, Portugal, Great Britain, France, and The Netherlands. These empires ventured into China, the East Indies, the West Indies, and the American continents. They introduced new products to Europe, such as tea, silk, and spices, but also fostered the creation of a massive slave trade, one of the most disgraceful commercial developments in our entire world history.

In the early 19th Century, at the end of the Napoleonic Wars, an era of modernization began to change many of the characteristics of the commercial world. Transportation across the seas was transformed with the use of steamships and over land with railroad networks. International trade increased. Production of goods became standardized. Rapid population growth increased the demand for commodities. Global agreements on monetary policy, trade policy, cultural exchange, political systems, and other human endeavors have multiplied over the years, resulting in our present enormous commercial world.

The pattern of colonization in Africa was unique in comparison to that of other parts of the world. In ancient times, people from Egypt, Greece, and Rome colonized the northern and western edges of Africa. Alexander the Great founded Alexandria, which has become one of the major city-seaports of the Mediterranean Sea. During the Middle Ages, North and East Africa were further colonized by people from Western Asia and then by Gothic Vandals. In the 7th century, the entire area was taken over by Arabs, who introduced their language and the Islamic religion.

Late in the 19th Century six European nations, Great Britain, France, Germany, Portugal, Belgium, and Italy, colonized nearly all of Africa. All that remained free were

Ethiopia and Liberia. The latter territory was settled primarily by black people from the United States, before the Civil War. The country became independent of the U.S. in 1847. The European occupations lasted until after World War II, finally completely ending in 1980.

Does globalization present a threat to our survival? Globalization is a combination of the commercial ventures and political structures of the world. Taken together, they are just short of functioning as a loosely connected world entity, but not as a planned, democratic, single federal government. The arrangement is a conglomeration of numerous political entities and the international relationships among them. The political aspect is reminiscent of the Articles of Confederation arranged by the former British colonies after the American Revolution. That arrangement worked badly. Finally, representatives from the thirteen confederated states met to attempt to modify the relationships among them. The attempt was not successful. Instead, the delegates made a major change. They wrote and approved the Constitution of the United States of America, organizing the semi-independent states into a single democratic federal republic.

However, the present world commercial arrangement is not undergoing major political changes. It continues to feature inequality and, on a wider scale, non-democratic world institutions, both commercial and political.

Most nations in the world are capitalistic in nature. Both capitalism and socialism have contributed to war, climate change, science, technology, and other instruments of the threat to survival. This has been shown in the expansionist efforts by the Soviet Union, China, and several smaller communist countries during the 20th Century, belatedly joining the world's capitalistic nations in this takeover venture.

No political-economic system in existence is close to perfect because humans are not perfect. It all comes down to the genetic differences, which will never go away, until and when *we all* go. The argument over the virtues of the free-market vs government may never end, except possibly in disaster.

Again, the primary purpose of business is to buy and sell goods and/or services. The financial bulk of the commercial world is formed by large corporations. Typically, they have personnel groups that include shareholders, managers, and employees. Their goal is to have enough income to remain in business, make a profit, and continue to expand.

How does business specifically contribute as a threat to our survival? The answer lies largely with the existential purposes of the various types of enterprise. For example, the armament industry is an obvious contributor to the world conflicts, as manufacturers and sellers of weapons of all types, plus military equipment, and other products that can be modified or adapted for military use. During World War II, many industries converted most of their production to support the war effort, producing military vehicles, ships, and airplanes, instead of cars and trucks for civilian use. The textile industry produced uniforms. Some foods were rationed, as were gasoline and other oil products.

As a practical matter, most businesses do not want war. Especially for companies that operate globally, war hinders business. Many companies will lose customers if they happen to be on the wrong side during a conflict. A company that tries to deal with customers in an enemy country runs the risk of being prosecuted for treason.

In peacetime, oil and gas producers, chemical manufacturers, and other companies may contribute some degree of threat, as a biproduct of their primary products, producing various pollutants that find their way into the soil, waterways, and the atmosphere. Some of these products contribute to climate change as well. In that case, a conflict may arise when the governments in countries where such companies exist attempt to regulate the emission of undesirable biproducts. The companies may resist either by lobbying the government or by claiming that they self-regulate. The latter is not likely to be a totally honest answer.

Business entities vary considerably in their methods of sharing revenue. In recent years particularly, very large shareholders may exert undue influence on company policy and garner very high monetary rewards by taking a large portion of profits, often at the expense of employee benefits. This has been made easier in recent years by the breaking of worker unions, so that the employees, who have little or no individual influence also have little group influence on company policy regarding salaries, a protective environment, and retirement funding. These selfish actions are a direct cause of economic inequality.

The business world embarked on a highly successful venture to influence politics in the United States when a lawyer named Lewis Powell, shortly before his appointment by President Nixon to the Supreme Court, wrote a memorandum to the U.S. Chamber of Commerce in 1971 stating the need for business organizations to contribute substantial funds toward influencing political movements sympathetic to conservative causes. The suggestion met with enthusiastic approval, affirming that freedom to contribute money to political causes was equivalent to free speech, as a First Amendment right. The Supreme Court agreed, eventually leading to the best- known decision on the issue, in a case called Citizens United v. Federal Election

Commission. Citizens United was a non-profit business organization claiming the right to advertise a film criticizing Hillary Clinton, a candidate for president, within 30 days of the 2008 Democratic primaries and with no strings attached. The Supreme Court agreed with Citizens United.

The decision encouraged mostly conservative non-profit organizations that already contributed money to political advertising increased their contributions to larger amounts. The Citizens United decision ruled that political speech was included in the definition of freedom of speech. The public was entitled to hear and read political speech under the same umbrella as any other type of speech. It was stipulated that contributions to candidates for office would not be permitted. Contributions could only be made on issues, through political action committees. No limits were placed on the amount of money contributed, and the source of the contributions could be kept secret. This arrangement is far removed from democratic principles. The source of *real* free speech is not hidden; it is a real live person.

There has been a great deal of complaint against the Citizens United verdict as expressed by the court majority. Justice John Paul Stevens wrote a minority dissent based upon the majority's lack of legal restraint. I doubt that the verdict was justified, but its real meaning is that the decision falls under the much larger realm of the nature of inequality: The decision meant that corporations were to be considered as persons and could contribute *unlimited* amounts of money to political campaigns, through non-profit super political action committees, *without* divulging the names of the original contributors. Money became speech, and the decision created massive inequality among voters and politically active entities.

Jim Hightower is a former Agriculture Commissioner

(1983–1991) in the state of Texas and easily qualifies as America's Number One Populist of the good kind. He publishes a newsletter, The Hightower Lowdown, which takes to task America's corporate giants that make tons of money by stomping on the heads of customers, working employees, small struggling businesses, and/or the unemployed. The Lowdown is particularly concerned with issues like "The State of the Plate" that asks the question: "Who Controls Dinner?" More on this in the essay on agriculture. The Lowdown is now a digital publication, called Substack newsletter. It is also a blogging service.

The Lowdown takes on other issues besides inequalities in agriculture, such as union busting by the giants in the energy businesses, the sad state of child-care in the U.S.A. as compared to other advanced economy countries, the continuing disappearance of local newspapers, the neglect of safe water delivery systems in some of our states and towns, and many other failures that subtract from our self-assessment of this country as exceptional in ways that deserve our pride.

An especially disturbing vital issue in American history is the longstanding assault against the labor movement by the rich and powerful in this country that essentially started *before* we became a country, in the Jamestown settlement founded in Virginia in 1607. The settlement's purpose was to start an extraction and shipping industry. However, the English "gentlemen" who came to Virginia were not much interested in doing work, a problem that was solved by importing men from Poland who would work for very little money. A problem that arose, when King James I permitted the formation of a local government, was denial of citizenship rights to the Polish workers because they were not English and had no property. Forget it, said the group of Poles; if we can't vote, we will not work. The potential loss of profits persuaded the owners to declare the Poles as Englishmen and therefore citizens. So much for the argument, sometimes made by corporate bigwigs, that unions were un-American.

Unfortunately, this story is not well-known and anti-unionism is a stronger force than it should be. Labor unions began forming in the middle of the 19th Century and went through many years of struggles to exist, despite the anti-union efforts of the companies, often supported by the federal and state governments. The election of Franklin Delano Roosevelt and the policies of the New Deal during the Great Depression enabled the eventual strengthening of the union movement, resulting in high levels of membership. However, anti-union conservatives gained control of Congress in 1946, followed by election of some Republican presidents, culminating in the 1980 election of virulently anti-union Ronald Reagan, a former Actors Union President!

The peak union membership percentage in the USA was 35% in 1954. In 1979 the total membership was 21 million. Private sector membership declined, but in 1960 public sector unions began to increase among federal, state, and local governments. Conservatives disliked the formation of unions in public service groups, muttering that the trend would lead to socialism, or worse. To top it off, Reagan may have overstepped his bounds in 1981, breaking the Professional Air Traffic Controllers Organization strike and firing all its members with little warning or opportunity for negotiation, a monstrous blow to public sector unions. The union was somewhat responsible as well.

I am pleased that my *alma mater*, Cornell University, has on its Ithaca, N.Y. campus the School of Industrial and Labor Relations, founded in 1945. Its mission is to prepare leaders and inform the public about national and international employment and labor policies. The school is one of the leading centers for labor issues education in the world.

12
Agriculture

Agriculture came into existence about ten thousand years ago when the hunter-gatherer groups living in the fertile crescent surrounding the Tigris and Euphrates Rivers apparently discovered that the food source plants in the area each year produced seeds as well as edible parts in a rather dependable manner. They learned to save some seed, replant, and produce a new crop the following season. They were able to settle nearby, reasonably confident that they could expect a crop each year. They built small villages and adopted a sedentary lifestyle. Their discoveries became known and were adapted by others in an ever-widening area. Several food plant species were domesticated, including wheat, barley, peas, lentils, chickpeas, and flax, as were animals, including cattle, sheep, goats, chickens, and turkeys.

Various locations in the world, similarly suitable for the domestication of various plants and animals, enabled hunter-gatherer groups to begin agricultural production and develop new civilizations. In the next several thousand years, various crops and animals were domesticated in China, Turkey, Greece, India, Africa, and the Americas.

Many years later, the plant species of the world that had been discovered and adapted in this way were grouped into eight centers of origin by the famed Russian geneticist and plant breeder, Nicolai I. Vavilov. His list of eight centers was modified by others in later years. Unfortunately, Vavilov ran afoul of dictator Josef Stalin, who was science-ignorant and easily hoodwinked by the phony scientist T. D. Lysenko. The latter discounted genetics as the source of variation that formed the basis for plant breeding. He claimed that crop plants responded primarily to changes in the environment, changes that became permanent. To safeguard his position Lysenko persuaded Stalin to send Vavilov to prison, where he died in 1943.

Lysenkoism remained acceptable to the Communist Party in the Soviet Union, with negative effects on agricultural production. People who rejected Lysenkoism were jailed or killed in most Communist countries. Eventually, the belief system essentially disappeared when its uselessness became almost universally acknowledged. Years later, it was discovered that certain environmental effects can affect human DNA directly, changing gene expression leading to generational changes. This phenomenon is substantially different from the crude effects claimed by Lysenko.

Because of the pioneering work by Vavilov and other scientists, particularly with plants, agriculture helped lead the way for all the other trappings of modern civilization to follow: a population explosion, industrialization, belief systems, science, technology, nation states and their governments, climate change, capitalism, socialism, endless conflict, and massive inequality that were probably little known in pre-history. Looking at this list, one can be overwhelmed, impressed, horrified, or all three.

How can agriculture be considered as a threat to our survival? Agriculture is the principal contributor of the food and drink that keep us alive, as well as materials for clothing fabrics, industrial crops such as rubber producing plants, and ornamental crops including flowers, shrubs, lawns, and trees. However, various chemicals are used for fertilizer or for protecting crops from insects and diseases. Those chemicals can also have deleterious effects on our health when they remain in the soil or get into the air and into excess water that escapes into streams and rivers.

There are other factors. Among the most valuable resources on the planet are the rain forests in the Amazon Basin countries of Brazil, Chile, and Argentina, plus Costa Rica, Congo, Alaska, Papua-New Guinea, Sri Lanka, and Malaysia. They all serve a vital function as carbon sinks, i.e., they store carbon that otherwise would be released into the air and contribute substantially to global warming. Unfortunately, many of the rain forests have been targeted for partial clearance to produce lumber, graze cattle, plant cereals and other crops, and for rubber tree plantations. These actions enlarge the world's carbon footprint, accelerating the pace of global warming, which already may be on a path to create a hot planet from which there may be no return.

It is particularly distressing that the former president of Brazil, Jair Bolsonaro, a climate change denier, had invited cattle ranchers and lumber producers to destroy part of the Amazon rain forest, which is the largest in the world. On the Pacific side of the world, the government of Papua-New Guinea has allowed logging companies free rein in destroying the country's forest. As noted above, other countries have also acquired badly needed income by allowing forest destruction, despite having signed the Paris Agreement to reduce the world's carbon footprint.

The United States supported the Paris Agreement of 2015, until the unfortunate election of Donald Trump as president in 2016. He is also a climate change denier, based on his incorrect gut feeling that climate change is a hoax. He ignored the extensive evidence accumulated over many years that global warming is real, created by humans, and increasingly a global threat. In June 2017, Trump announced that the United States would withdraw from the Agreement in three years, the minimum time to activate a withdrawal. The termination took effect on November 4, 2020, one day after the Presidential election! Fortunately, with the election of Joe Biden as President, the United States has rejoined the agreement.

The possible harmful effects of agricultural activities should be balanced with its benefits. Agriculture produces food, fabric, wood products, natural rubber, and beauty, thus accounting for more of our basic needs and desires than most other human activities.

The science of genetics was born in 1865 in Brno, Austria (now in the Czech Republic) in an abbey garden, where Friar Gregor Mendel grew garden peas. He studied the inheritance of seed color and several other traits in the offspring of crosses between pairs of parent plants. He was then able to formulate the laws of inheritance, which were later named after him. The science of genetics and its myriad applications in plant and animal breeding also became useful in medical and social studies, including at the molecular level, significantly expanding the bounds of human knowledge.

Agriculture led the way in pursuit of many biological knowledges, in addition to the production of consumable products, thus permitting the expansion of the world population, the building of cities, the creation of vast industries, and the opportunity to make constant choices, both good and bad. The last phrase in that last sentence brings up an interesting controversy about the very existence of agriculture. Was the dawn of agriculture a great step forward for mankind or, as claimed by author Jared Diamond, mankind's worst blunder?

It certainly was a great step forward, in the sense that it was the 'big bang' of human civilization: leading to a tremendous explosion of new things to eat, not just grains, but many other edible species; a new way to shelter; less need to wander; and an opportunity to meet new people and share ideas. There should be no surprise that not all the ideas were good ones, because they never are. We were and are genetically varied. We do things that are good, bad, and neutral. Because of evolution and genetics, we continue to develop tremendous variability in our species for every trait that defines us.

So, were the changes made beginning ten millennia ago a great blunder? Of course not. A blunder is defined as a stupid or careless mistake when confronted with a meaningful choice, specifically that those involved in changing from hunting-gathering to planting-harvesting knew what they were doing in terms of eventual consequences. Not at all. They did not have a meaningful choice.

Were people better off before the agricultural revolutions in the Fertile Crescent, the New World, East Asia, and Africa? It really doesn't matter, because we are not going back, at least not voluntarily. Of course, the Bushmen and other modern hunter-gatherers might say that they are better off than the rest of us and could possibly make an argument why that is so. But it would be pointless to carry out such an argument.

Dr. Diamond did point out that our species is in trouble, that choices, nutritional and others, have been made that have done harm to many of us. Some nutritionally unwise choices have been made by people for themselves, even though nutritional information exists and should be followed. Unfortunately, other choices are also being made by the few for the many who are often poor. Many poor people feel that they must eat as cheaply as possible. This often means "fast food", which can be relatively unhealthful. These are problems of genetic behavior traits and economic inequality.

The importance of agriculture is that it is the easily labeled beginning of revolutionary change in the planet's destiny, featuring the explosive development of all the institutions described in this writing. I think it mislabels the event to call it mankind's greatest blunder. Considering the capacity of the human brain, it is difficult to imagine any other destiny more likely to have happened. It certainly is impossible to believe or expect that humans would be hunter-gatherers forever. And yet the change took so long to happen, hundreds of thousands of years. Perhaps humans tried agriculture earlier, but it didn't work out. Why not? No one knows because we were not there, and there may be no evidence left to be discovered so many millennia later.

Another consideration. The hunter-gatherers that are still with us are highly conservative. Conservatism is a very powerful belief system. Galileo discovered that. Consider the immensely long periods between significant inventions in ancient days. And yet the discovery of agriculture did occur and was independently repeated all over the world within a rather short time frame. We may have to make similar meaningful, major changes in our life pursuit if we wish to remain here at least for another ten thousand years.

Agriculture has been closely associated with one institution that might be considered for the title of humankind's greatest blunder, also worst crime, and most shameful institution ever. That institution, of course, is slavery, and it is very old, possibly originating in the first city-state, Mesopotamia, in the year 6800 BCE. In the early days, men captured in battle became slaves of the victors. They worked on ships, fought and died as gladiators, or labored in mines and agricultural fields. Those fates were also suffered by debtors, captured criminals, and others of depressed standing.

From its early start, slavery became a worldwide institution and a major source of wealth in Europe, Asia, and Africa for centuries. Slavery slithered into the New World in the 16th Century. The Spanish invaders, after the "discovery" by Columbus, fought against the native populations, forcing them into labor as well as killing them off in battle and transmitting smallpox and other diseases, of which the Spaniards were carriers, but which were utterly devastating to the natives.

British, French, and Dutch colonists took possession of what is now the eastern United States and Canada, starting in 1619. At the time of the Revolution in the 1770's, slavery existed in all the colonies. It was especially widespread in the south, in colonies run by wealthy plantation owners, who preferred to believe that the backbreaking work in the cotton, tobacco, indigo, and sugar cane fields could only be handled by black people. Slavery became a permanent institution, because children of slaves were also slaves. It was, of course, an exceedingly cruel, violent institution, featuring whipping as punishment, rape of slave women by the owners and overseers, trading of slaves, often breaking up families, and other equally monstrous treatments and punishments.

Slavery in the United States legally ended with the Civil War, partially with President Lincoln's Emancipation Proclamation in 1863, and completely with the 13th Amendment to the Constitution, in 1865. The twelve-year period following the war was called the period of reconstruction. Its purpose was to encourage the freed black people to assume the rights of citizenship. Unfortunately, opposition by the southern whites prevented true freedom for black people. Then, when reconstruction ended, the southern states immediately enacted laws that prevented black people from voting, restricted use of public facilities such as water fountains and public bathrooms, created segregated schools and churches, and forced blacks to sit in the back seats of buses. Also, they could not sit at a table in a restaurant, but only order to take out. Those restrictions can only be described as hateful and stupid. One of the worst offenses, of course, was the vicious practice of the lynching of black males, a crime led by the Ku Klux Klan. Lynching was done primarily as an expression of hatred, often for trivial reasons or for no reason at all.

Finally, in the 1960s, Congress passed civil rights legislation that mandated the right to vote for all, desegregated public schools and other public accommodations, strengthened the Civil Rights Commission, prohibited discrimination in federally assisted programs, and provided for equal employment opportunity. A horrible consequence of these changes was the assassination of Martin Luther King Jr., the most prominent leader of the civil rights movement.

The civil rights and voting rights legislation led to a series of strategies by the Republican Party, employing language that appealed to southern Democrats, particularly by implying that by switching from the Democratic Party to the Republican Party, in national, state, and local elections, policies could be enacted that would be more friendly to white segregationists and less to black and brown voters. The language emphasized states' rights and opposition to school bussing. The strategy began during Senator Barry Goldwater's presidential campaign and continued through the campaigns of Richard Nixon and Ronald Reagan. The Republicans no longer call themselves the party of Lincoln. Surprise!

In the 2016 presidential election, Donald Trump, with his ridiculous "Make America Great Again" campaigning, dug deeply into the wrong side of racial politics. He eventually paid the price with defeat in his 2020 campaign for a second term. His 'contribution' to agriculture included reducing postal services for rural areas, eliminating rural development as a federal mission, funding and supporting large agribusiness companies at the expense of small and medium size farms, allowing meatpackers to speed up butchering lines that endangered line workers, and hampering research.

As noted previously, Jim Hightower is a former Agriculture Commissioner in the state of Texas. He publishes a newsletter, The Hightower Lowdown, (now a digital publication) which takes to task America's corporate giants that make tons of money by stomping on the heads of customers, working employees, small struggling businesses, and the unemployed. One issue of the Lowdown was particularly concerned with "The State of the Plate" that asks the question: "Who Controls Dinner?" His response is suggested in the last sentence of the previous paragraph.

The farms and ranches that produce the major crops: grain species that feed cattle and fowl; meat products; sugar beets, a major source of sugar; cotton; rice, and wool-bearing sheep, are mostly under the thumb of large corporations that purchase, refine, and sell the products. However, horticultural crops: vegetables, fruits, flowers, and ornamentals are grown by farmers less dependent on the giant corporations. The participants are growers, shippers, and grower-shippers, each group consisting of small, mid-size, and large companies.

Agriculture has been a pioneer and major partner to science and technology, based on the extensive use of crop plants and animals as experimental subjects for studies in genetics and breeding, physiology, pathology, insect vectors and pests, virology, growth habits, nutritional value, and other plant characteristics. Research organizations that conduct agricultural research include the U.S. Department of Agriculture through its Agricultural Research Service headquartered in Beltsville, Maryland. In addition, there are five geographic areas, each containing several local federal research stations.

In the educational system, State Universities have varying numbers of research departments performing agriculturally related research, oriented to solving problems in each state for vegetables, fruits, cereals, forage species, economics, plant pathology, plant physiology, genetics and breeding, and animal needs.

The pioneering experimental methods developed in agriculture have also been found to be useful in medical and human genetics studies. These methods have been largely responsible for the great advances in those fields.

13

Inequality

"We hold these truths to be self-evident, that all men are created equal, that they are endowed by their Creator with certain unalienable rights, that among these are life, liberty and the pursuit of happiness."

These are treasured words in the Declaration of Independence and had been accepted as true by Americans, until a grave political and humanitarian need for corrections was brought into view: All men meant all men of property, it meant white men, and it meant no women. This definition changed little by little until the middle of the twentieth century, when the denial of voting and other rights was finally completely abrogated by civil rights and voting rights laws signed by President Lyndon Johnson, a Texas Democrat, thus bringing us closer to equality. Unfortunately, efforts are still being made in many states by the Republican Party to set up barriers, meant for people of color, to make voting more difficult. It may seem strange that the "party of Lincoln" is responsible for erecting such barriers, until we ask the question. Who are these people? They are the former southern Democrats and their descendants. They had been a segregationist blot on the Democratic Party for many years, until Johnson's actions. Then they were further persuaded by the likes of Goldwater, Nixon, and Reagan to become Republicans and all would be okay. A pox on their houses!

Of course, many kinds of inequality are independent of law. They are inherited. Inequality is an unavoidable consequence of the variability of traits borne by the human population of the world. It is unavoidable because genetics and evolution have made it so. Except for identical twins, no two people are exactly alike: physically, mentally, emotionally, or behaviorally, nor can they be made alike by manipulating their environments. Inequality, with its many faces, is a consequence of genetic variation built up to great proportions by mutation, evolutionary and personal selection, movement of humans to all parts of the earth, and an almost constant increase of the population since our earliest days.

The phenotypic (observable) distribution of quantitative human traits can be displayed with a bell-shaped curve. (See Essay One) The left end of such a curve is at zero persons. Moving to the right, it climbs gradually, then steeply to the top of the curve, turns downward, then steeply down, slowing again until reaching zero. Examples of human quantitatively inherited traits distributed in this way are: height, skin color, intelligence, behavioral traits, and personality traits. For any of these characteristics, the number of persons at locations on the curve is indicated by the height of a vertical line from the base to the curve. The distribution of the lines is modified by the effect of the environment, which smooths the curve.

For example, let's consider individual heights of a large randomly selected group of human females of a similar age. The shortness of a vertical line near the left end shows very few persons at that height. The number of persons increases as the curve height increases from left to right, then decreases down to the other end. For height of females, therefore, the value from left to right goes from very short to medium height at the top of the curve, then decreases to very tall on the right end of the curve. The same changes would occur for a similar grouping of males and other variable groups.

To get a better picture of our genetic variability and its relationship to the concept of inequality, we would need to know, or at least estimate, the number of both qualitative and quantitative genes in the human genome. We must go beyond the bell curve and consider the number of possible genotypic combinations. That number is very large: over 70 quadrillion based on the number of protein coding DNA genes (about 20,000) plus the number of non-coding DNA and RNA genes, which total many thousands more. The number of possible combinations easily exceeds the number of people living now, eight billion, and even the total number who have ever been on earth, over 100 billion. There are no two identical genotypes in the world population (except identical twins, as noted).

However, there are groups of similar genotypes: perhaps among various groups that have similar interests and/or ideas. These might be religious groups, political groups, sports fans, actors, business leaders, teachers, and others. People in the same genotypical bracket may or may not know each other. Yet they may all have similar sets of genes for behavior, personality, intelligence, height, and other traits, and may, under certain circumstances, join the same political circle, religion, club, or some other group or organization. Further, they may react in the same way to a news story: pleased, angry, amused, or indifferent; vote for the same presidential candidate; harbor prejudice against, or for, the same kinds of people, and so on.

Inequality among people is certainly a very old characteristic, although it is likely that the earliest small human groups had less genetic variation. When enough generations passed, the population increased, mutation and selection did their work in the creation of more genetic differences, and eventually, inherited inequality asserted itself more strongly. Hunting and gathering tasks may have been shared by the men and women of an early wandering group. This certainly changed about ten millennia ago with the conversion to a sedentary life, based on agriculture. At that time, a broader inequality would then have become a hallmark of humanity.

Behavior and personality genes influence the character of persons carrying them. Therefore, they contribute to various inequalities depending upon the distribution of the genes in different persons. Inherent human inequalities are numerous, illustrating the genetic differences among people. The genetic effects might be somewhat influenced by economics, social standing, education, location, family structure, political structure, religion, ethnicity, gender, and other environmental effects that modify the relationships among and within groups. In democratic societies, efforts are often made to minimize many of the unfair inequalities. These efforts are delicately balanced and may change for better or worse when the political leadership changes.

Inherited inequalities change in a meaningful way only through mutation and selection. These include variations in intelligence, many behavioral and personality traits, reaction to genetic diseases and disorders, and physical traits. These traits are genetically built in. Various types of economic inequality are based upon the existence of these non-changeable differences.

In addition, social prejudices lead to deliberate imposition of discriminatory actions against minority groups based upon religion, politics, skin color, ethnicity, gender, and other characteristics of varying importance, associated with appearance, accent, or style of dress. For many groups of people, such discrimination pushes them farther down the road to greater poverty, loss of educational and residential opportunities, and inability in general to reap the fruits of our modern Garden of Eden.

As a result of our genetic differences, we have lived in a world of inequalities, now and during the many past millennia of our sojourn here. They add up to a figurative steamroller, crushing the minds, hearts, and spirits of too many of the eight billion inhabitants of the planet. The purpose of the explorations here is to try to determine if there is a way out of this path to disaster, along with those discussed in the other essays.

Unlike genetic inequalities, those created by society, deliberately or through neglect, can be reduced by societal action designed for the purpose. The positive actions should include the provision of financial support for schools in poorer sections of cities and in rural areas, alleviating the monstrous differences in corporate income of the executives compared to the average employees, and making available adequate treatment centers for poor people, particularly for minorities (black, Latinx, and native Americans in the United States, and other similarly deprived groups in other parts of the world). As of this writing in mid-2023, the COVID pandemic active over the entire world from 2019, has subsided substantially. It has been an onslaught on human health not seen since influenza covered the world in 1918–20. Vaccines have been developed and have been effective in protecting those who are vaccinated and received subsequent booster shots. As usual, the same groups that have been neglected for other needs have suffered similar neglectful treatment. We don't know how long it will be until poor folks, inhabitants of rural areas, and people of color will be protected on an equal basis with more fortunate people.

Is it possible to make improvements in our world to minimize the level of inequality and thus to improve the living conditions of those who, currently, are suffering the various consequences of their plight? If so, it will be necessary to make improvements in our world society to ameliorate the genetically based inequalities.

What sort of improvements can we make? One of the major inequalities of life on this planet derives from the planet itself, which has presented us a range of environments, some of which are friendly to human life and others which are highly unfriendly. People must deal with extreme heat in the deserts; extreme cold, snow, and ice in far northern and southern climates, and in high mountains; heavy rains in the tropical jungles; and infertile soils in many environments.

The inequalities of location are further complicated by poverty. The owners of large energy companies refuse to admit that which they know to be true: the earth is warming, climate change is real, and it is a product of human activity. If we fail to react properly by replacing fossil fuels more rapidly with strongly mitigating measures, we will bring on a hot earth much too quickly. We will all suffer to some degree, but the poor will suffer more than any other group.

These situations can be aggravated if a nation tries to

improve its lot by invading a more prosperous neighboring country. Similarly, within a country, one group may attack another with the same purpose in mind. In Rwanda, over 100 days in 1994, the Hutu ethnic group committed genocide against the Tutsi group, a terrible happening allowed to take place by other countries that could have prevented the extensive murdering.

Our treatment of the planet discussed in the first essay can also be considered a form of inequality. An article published April 18, 2022 in The New Yorker, entitled "Testing the Waters" by Elizabeth Kolbert, discusses several legal issues, conflicts, and lawsuits regarding the question: Does nature have rights? The question can be asked regarding the fate of streams, lakes, valleys, swamps, beaches, trees, mountains, hills, plant groups, etc. These natural phenomena can be compared to non-living entities, such as corporations, which *are* given rights in legal confrontations.

During the years following the first invasions of the American continents by intruders from Europe, many of the native tribes did not understand the need of the white settlers to own land and other kinds of property. The tribes apparently had minimal desires for personal ownership, limited to their clothing, their tools and weapons, and their homes for as long as they lived. The land on which the tribe lived belonged to the tribe. These niceties meant little or nothing to the invading Europeans and soon-to-become Americans. The land was ripe for the taking even though the tribes had been living there for hundreds or thousands of years.

In 1622, shortly after the arrival of English settlers in Jamestown, Virginia, they were attacked by the Powhatan Tribe, in a massacre that encouraged the English government to retaliate by attacking the Indians and confiscating their land. This was followed by the Pequot War between the British and tribes in the vicinity of Massachusetts Bay, and battles between Dutch settlers and several tribes in New York. Also, French settlers fought the Iroquois over fur trade dominance around the Great Lakes.

King Philip's war (1675–6) began after Wampanoag bands tired of their dependence on the Puritans, and they attacked colonies and militia camps in Massachusetts and Rhode Island. Queen Anne's War (1702–1713) was a long war between French and English colonists, each with Indian allies. The French moved into the Ohio River Valley during the period 1754 to 1763 and fought with Britain for control of North America. Both sides had Indian allies. Known as the French and Indian War, it ended with the signing of the Treaty of Paris.

When the Revolutionary War broke out in 1775, the tribes had to choose to fight on the side of Americans or with the British or try to remain neutral. In the end their choices did not matter: They were left out of the peace talks in 1783 and lost more of their lands.

Battles between the Americans and the tribes after the Revolution lasted until the late 19th Century. One of the best known was the Battle of Tippecanoe in 1811. The American forces led by William Henry Harrison won the conflict. Harrison later ran for the U.S. presidency at the age of 68 with the slogan "Tippecanoe and Tyler Too". Unfortunately, he contracted pneumonia and died after one month in office. John Tyler assumed the presidency.

Eventually, the fate of the U.S., Britain, France, the Netherlands, and Spain, each of which had varying ambitions for expansion in the New World, reached landmark proportions in 1803 when President Thomas Jefferson purchased the Louisiana Territory from Napoleon, almost doubling the size of the United States.

In the ensuing years, the government's goal was, broadly speaking, to herd the tribes onto reservations west of the Mississippi River and strongly encourage them to become farmers and Christians. After three wars, the Seminoles agreed to be settled in the Oklahoma Territory. President Andrew Jackson's personal hatred of the native peoples impelled him to sign the Indian Removal Act in 1838, forcing the removal of 15,000 Cherokee from their southeastern home and walk 1200 miles to the west. The route was named the Trail of Tears because over 3000 lives were lost on the long, extremely arduous trip.

Earlier, in 1832, Chief Black Hawk led Sauk and Fox Indians to Illinois to reclaim their lands, but they were greatly outnumbered, resulting in a disaster for the Indians. In 1876, General George Armstrong Custer, and 600 men were all killed by 3000 Sioux and Cheyenne in the Little Bighorn Valley.

Finally, the last major conflict was the Wounded Knee Massacre in 1890 when the U.S Army surrounded a group of Ghost Dancers and slaughtered 150 Indians. By the early 20th Century, the wars had ended, but at tremendous, accumulated costs. Overall, the Indians helped the British colonials to survive in the new lands and gain their independence. Later, however, they were forced to cede large amounts of land to the newly independent Americans. Many thousands of their lives were lost to war, disease, and famine, and the Indian way of life was essentially gone.

The ultimate state of inequality is slavery. Most slaves own nothing, not even their own bodies. There were several ways by which human beings first became slaves. One was to be captured in battle and thus lose self-ownership. Those unfortunate men might then have become gladiators, boat rowers, miners, farmhands, artisans, or household servants. The same punishment was also applied to debtors. These forms of slavery were part of the earliest known instances of slave containing societies that came along after the hunter gatherer ancient world essentially disappeared.

Slavery may have originated in Sumer about twenty centuries BCE and was practiced in various forms in Ancient Egypt, Judah, Greece, and Rome, continuing through the Middle Ages into many countries of the modern world, up to and including the present time. We are most familiar with chattel slavery, in which slaves are characterized as personal property, especially during the period of colonialism and the pre-Civil War United States. Since the Civil War, we have been struggling with the fall-out from that distressing period. Prejudice against black people, especially in the southern states, still exists. It is manifested by laws making it more difficult to vote, especially for black citizens.

Slavery also occurs under different forms and names. Indenture refers to labor by which a person pledges to work and pay off a loan. Slavery also includes child labor, child soldiers, forced marriages, and other situations that occur through coercion. Some people believe that military drafts and even taxation should be considered slavery.

Slavery continues to be a blight on planet Earth. The number of slaves today is estimated at between 12 million and almost 40 million people. It is intolerable that slavery still exists. **All slavery must be eliminated.**

14

Politics: from Liberal to Conservative

Politics—A word that has a variety of definitions and can be applied to the relationships among individuals, groups, cities, and other governmental entities with both positive and negative meanings that will apply to all aspects of the discussions that follow here. The words liberalism and conservatism both have a variety of definitions. Liberals may be labeled as left wing, progressive, radical, socialist, and communist. Conservatives may be called right wing, libertarian, radical, and fascist. The defining words usually depend upon the attitude of the one who selects the words.

Consider the following set of descriptive terminologies for politics:

a. The activities associated with the governance of a country or other area, especially the debate or conflict among individuals or parties having or hoping to achieve power.
b. The activities of governments concerning the

political relations between countries.

c. The academic studies of government and the state.
d. Activities within an organization that are aimed at improving just one person's status—the leader. Such activities are generally considered to be devious and divisive.
e. A particular set of political beliefs or principles.
f. The political assumptions or principles relating to or inherent in a society or section of a society.

These descriptive statements nicely sum up the variability employed in defining politics. Each one can be further described as having either beneficial or harmful effects on some or all the people involved in the appropriate relationship. For example, democratic countries have two or more political parties of either a liberal or conservative nature, and decisions range from fair and acceptable to unfair and unacceptable to one or more of the parties involved. A common reason for this sort of division is the imposition of taxes, which are accepted by some and disliked by others.

Among countries, political relations are international in scope and may involve two or more neighbors or most of the world. Relationships may develop as a dispute over land between two countries or a series of disagreements involving several antagonists that might eventually go to international organizations like the United Nations or NATO for settlement.

Academic studies of government and the state are found at secondary schools and universities worldwide and are meant to foster understanding of national and international political relationships.

Political systems are belief systems, meaning that real, well-thought-out evidence might not be used to support beliefs. Within a country, relations among parties can be benevolent or antagonistic in nature, depending on the beliefs held by each party. Among countries, similar relationships will hold, but bad relations may lead to war, far beyond the bad feelings that may exist among parties within countries and/or among individual persons—unless, of course, disagreements end up as personal violence or a civil war.

In a broad sense, the word liberal refers to people who welcome change and are open to new ideas. A liberal person or party will consider making governance changes and is usually willing to support them and pay the costs. A conservative person is less amenable to these changes, an attitude often based on dislike of ensuing tax increases that diminish profits.

If one examines the philosophies that define our existence on the planet Earth, the exercise will enlighten us on the genetic basis for our differences and similarities—and reinforce the theme of this book: As discussed earlier we are the beneficiaries and the victims of our genes, which are largely responsible for the threat to our survival.

Consider liberalism. If we study the definition of the term, we find that it has changed multiple times since the Age of the Enlightenment in the 17th and 18th Centuries. These changes have been dependent upon the institutional makeup of the nations: their governments, peoples, histories, religions, educational levels, financial states, and paths of change over the centuries. Each of the national institutions became adapted in some way to current thought on those issues during the centuries of conscious recognition of the various concepts, especially the relationship between the individual and the state regarding rights and obligations.

The earliest definitions of liberalism were developed in the late 17th century by two English philosophers, John

Locke and Thomas Hobbes, both of whom approved the following concepts:

Belief in equality and individual liberty

Support of private property and individual rights

Support of the idea of limited constitutional government

Recognizing the importance of pluralism, tolerance, autonomy, bodily integrity, and consent.

Hobbes advocated a strong monarchy, claiming that only an absolute sovereign would be able to sustain the needed security. Locke, however, believed that the royal government requires consent from the governed, preventing the sovereign from becoming a tyrant, and that the people have the right to overthrow a tyrant. He did not accept the existence of divine right. He also originated the idea of separation of church and state. Politician and poet John Milton also advocated disestablishment, and the equality of all humans. As assistant to Oliver Cromwell, he was one of the first to advocate freedom of speech.

Regarding the powers of government, the American liberal theorist and fourth President James Madison and the French philosopher Baron de Montesquieu advocated separation of powers among the executive, legislative, and judicial branches. This separation was instituted from the beginning of the newly formed United States government.

From the 19th Century to the present, the concept of liberalism took several different meanings in deference to the variation among political, religious, employment, and philosophical groups. The differences are due to the ways in which people view the meaning and importance of various cultural terms and ideas. These include individuality, adversariness, rationality, democracy, republicanism, and other descriptive terms.

Liberalism is a political doctrine that emphasizes protection and enhancement of the freedom of the individual as the central concern of politics. Liberals believe government has an additional role. It must protect individuals from harm by others and refrain from posing a threat to their liberty. A thought separation occurs here. Libertarians (neo-classic liberals) believe this is all that government should contribute. Most liberals however, since the late 19th Century, argue that more is required: the government should promote freedom beyond the basic requirement. In the United States, modern liberalism is associated with the welfare state policies of the New Deal created by the Democratic Party of President Franklin D. Roosevelt to end the Great Depression. In much of Europe, limited government and laissez-faire economic policy are the important watchwords.

In the 18th and 19th Centuries, the concepts of liberalism were clouded by the founders' fear of total participation by the citizens in the American and European countries. Participants in governments included only propertied men in Europe and only propertied white men in the United States. Gradually, participation was widened, eliminating the criterion of property ownership. In the United States, universal white male suffrage became law in 1860. This was augmented by black male participation in 1870. Female suffrage in America was made law in 1920 after years of struggle. These advances in participation were partially made possible as the results of wars. The overall changes in America and Europe were made possible by the outcomes of a series of conflicts. In England they were in response to the English Civil Wars (1642–1651) and the Glorious Revolution (1688). In the United States, it was the American Revolution (1775–1783) and the Civil War (1860–1864). In France, it was the French Revolution

(1789). The specific political ideas operating varied among the countries.

Conservatives respect institutions that have existed and changed slowly over long periods of time. Most specifically, they demand that the government should be the servant of the people and not the master. Conservatism began to develop as a specific political movement in the late 18th Century. The French Revolution in 1789 spawned political and religious upheavals leading to a reactionary response by Vicomte de Chateaubriand and others, during the restoration of the Bourbon monarchy. Other countries, including the United States and the United Kingdom, approached the concept of conservatism in various ways, depending on their definitions or understandings of conservatism, liberalism, libertarianism, and other related terms.

It can be edifying and interesting to list the many forms of conservative belief as adopted over time and in various places around the world:

- To prefer the familiar to the unknown, the tried to the untried, the actual to the possible, the convenient to the perfect etc.
- To promote norms that could stand timelessly
- A philosophy of human imperfection
- Respect for authority and religious values
- Constitutionalism, paleolibertarianism, small government conservatism, Christian libertarianism, free trade, opposition to national banks, and opposition to business regulations
- Prudence in government spending and debt accumulation
- More emphasis on national interests and upholding cultural and ethnic identity
- Government enforcement of traditional values

- Upholding traditional family structures and social and religious values
- Progressive and wise conservatism
- Authoritarian and reactionary conservatism

It is possible to define personal conservatism as a form of selfishness. A different personal assessment is that it is simply a fear of being replaced.

Politics can be labeled with numerous descriptive terms: fascinating, dull, necessary, unnecessary, espoused by honorable people, espoused by dishonorable people, and many other pairs of opposite meanings. Undoubtedly, politics will exist along with human beings and their differences.

As long as the world and its population survive, so will politics, which is, after all, a consequence of the existence of that population. The tremendous number of people and their differences will display, as always, a conglomeration of political differences among people, their families, their groups, and their cities, towns, and villages. As time goes by, the world portrait will continually change, for better or worse, as long as humanity exists.

This examination of politics requires an emphasis on the importance of discussing civics for the citizens of any political structure. Civics is defined as the rights and obligations of citizens in society. The subject is more likely to be taught in a country that is a democracy, where students are encouraged to respect democratic values and to practice them, when in society as full-fledged citizens.

We hope that someday all or most countries will value and practice democratic ideals, so that their citizens will engage them also. Unfortunately, in the United States at present, only about one-fourth of young, secondary students has been educated in civics and tested for civics proficiency in recent years.

15

History and Geography

We describe the historic stages of our planet Earth and our presence on its surface by recording what has happened. The period before history, beginning with the use of stone tools by early *Homo* species, is labeled pre-history. The pre-historic record is based on bones, teeth, and inorganic materials left unwittingly after the death of their possessors. This distinction stems from our human predilection to acquire a great deal of information about ourselves and to examine our past existence. Life began millions of years before we came along, and the planet itself was formed over four billion years earlier. The universe was born about 13.8 billion years ago from a small particle that was very heavy and very hot. It exploded with a "big bang". What existed before the big bang? Alas, we have no solid idea and may never be able to answer that question. It could be that there had been nothing, a void. Or there might have been a series of existences, each one possibly like our present one. Or not. Truly a mystery.

Let's return to the human point of view and the goals of this book. History can be a teacher, but only when we pay attention and learn useful lessons from events of the past, those that pass through our own lives, and from future events that may continue to occur for some unknown length of time. This is not as obvious and as simple as it may seem. Why? We tend to accept the meaning of events in terms of ideas and conclusions developed earlier, so that the events appear to confirm what we already believe, which could be right or wrong. We accumulate belief systems. This may stem from the common notion that history repeats itself. It really doesn't, because circumstances change, and the people involved in the earlier events die or become older and are replaced by new people. Many occurrences are similar and may seem repetitive, but others are not.

Consider a sequence of wars. For example, World War I was followed twenty-one years later by World War II. Adolf Hitler was an ordinary soldier in the German army during WWI. Furious with the terms of the armistice and with Germany's post-war leadership, he took up politics and joined the German Workers' Party, which morphed into the vicious Nazi Party. He became Germany's dictator in 1933, and soon proceeded to inflict WWII on the world.

WWI was fought in Western Europe and also in the Balkans, Africa, Italy, and the Pacific, and it lasted over four years (1914–1918). The United States entered the war in 1917 and participated for a year and a half until the war ended with an armistice. When the American troops landed in France, commanding officer General John Pershing announced: "Lafayette, we are here."

WWII was a six-year conflict, 1939–1945, and the weapons used were considerably more powerful than in WWI. The U.S. declared war on Japan December 8, 1941, the day after a Japanese bombing attack on the American fleet in Pearl Harbor, Hawaii, and declared war on Germany, Italy and three Balkan countries three days later. The war was fought on several fronts in the European-African theater and in the Pacific theater, and it ended with the explosions of two fission type atom bombs that killed many thousands of people in two Japanese cities. Nearly all were civilians. There are lessons to be learned from those happenings, but it is difficult to decide which examples will be truly helpful and heeded. One sure lesson is that a full-scale nuclear war would be totally disastrous. Ignoring that possibility might well leave us with a world that would be truly uninhabitable.

Claiming that history repeats itself may be a means to avoid thinking about the differences between one event and a second that follows at some future time. Those differences may be quite edifying and if studied carefully will disclose the non-repetitive aspects of the two occurrences, which might then allow them to be compared more meaningfully. One of the big differences was the almost total absence of trench warfare, a fixture in previous wars, during World War II.

History allows us to see the past, analyze it, and consider changing the way we do things to avoid past mistakes. It's important to understand the meaning of past events and the people involved in them and perhaps improve our interpretation of them, thus benefitting ourselves and hopefully the world. This is an ambitious goal and not easy to reach; we often get caught with ever more emotional assessments and make new mistakes as well as repeating some of the old ones. Wars may become more destructive, climate change may take a more damaging path, and the road to disaster might get shorter.

It may seem that the words history and war are very nearly synonymous. Certainly, a war is often an historical event. Would the end of war-making mean the end of history? A great significance of war is in its multiple characteristics: mobilization of large numbers of people, violence, death, injury, tremendous costs, property destruction, setting aside of peacetime activities, feelings of hatred toward people of the other side, and persistence of hard feelings beyond the end of the conflict; all of which, and more, make a war a landmark event. It may also set the stage for another conflict. War also fosters the appearance of new ideas, newly invented devices, and changes in the way people carry out their affairs. In the past, these were common outcomes. In the future, the outcome may be a destroyed human civilization and perhaps the disappearance of all life.

A world without war would be very much different from its present state. It is difficult to imagine the ways in which those differences would frame our behavior, our ambitions, our satisfactions—and our dissatisfactions. In the last essay, I will discuss my thoughts about the possible relationships in the absence of war, among people and among and within countries. and how people will think of each other and about themselves.

For the rest of this essay, I will continue discussing the passages of the various stages of history, starting with ancient history, the period immediately after pre-history. Pre-history is the time before humans developed means of recording historical events. Artifacts pointing to the existence of pre-history were found accidentally at first. Archeologists then began actively searching for further evidence: bones, tools, and other artifacts. Pre-history began with the invention and use of stone tools 3.3 million years ago.

Ancient history is usually designated as the period beginning with the adoption of writing, about nine thousand years ago, in the Early Bronze Age. Writing began with the use of symbolic pictures: Mesopotamian pictographs and cuneiform, Egyptian hieroglyphs, Proto-Elamite pictographs, to development of Phoenician and other alphabets. Writing systems were also found in Mesoamerica by 500 BCE. Chinese script has been dated to about 1200 BCE. The cuneiform system lasted for more than three millennia, into the second century CE.

The birth of agriculture took place about ten millennia ago. The wild plants that became crops included wheat, barley, flax, chickpeas (garbanzos), vetch, lentils, tree fruits, and grapes. Agriculture intensified in the early Bronze Age (about 3300 BCE). The word agriculture most commonly refers to planting and harvesting occurring in the lands surrounding the Mediterranean Sea in Europe, as well as in Africa, and in the Middle East. This period ended with the fall of the Western Roman Empire, while the Eastern Roman or Byzantine Empire survived and prospered.

The Middle Ages began in the Fifth Century and lasted until the Fifteenth Century. It was a period largely devoid of leadership, which had been provided by the Roman Empire. That leadership was not one of benevolence, which was not at all common in those days. It was based on the imposition of power, a universal phenomenon of human history. When the Empire collapsed, the closest substitute for leadership on the European continent was the Catholic Church. The Church supported art, music, education, architecture, and of course, warfare, especially The Crusades, ventures meant to defeat the Muslims and reclaim the Middle East for Christianity. There were eight major Crusade expeditions between the years 1096 and 1291. Most of their efforts failed, and the area has remained primarily Islamic and more recently, Judaic.

The next period is called Modern History, which

essentially began with the four voyages of Christopher Columbus to the New World in the late 15th Century. His expeditions were sponsored by the Spanish monarchs Ferdinand and Isabella. Modern History continued until the late 18th Century. Spanish, British, Dutch, Portuguese, and French explorers sailed across the Atlantic Ocean, and proceeded to plunder and/or settle in countries large and small, and then to create empires. In Europe, Modern History included the Renaissance and the Ages of Reason and Enlightenment, which were followed by the Industrial Revolution. Now we are in the third century of what we could call Present Times or Nowadays.

Let's turn back to the early histories of other parts of the world: eastern Asia, Australia, Oceania, and the American continents. Descendants of our Australopithecine ancestors in Africa evolved into early *Homo* species and then to *H. sapiens.* The earliest migrations were undertaken by the last human predecessor*, Homo erectus,* about two million years ago, into Eastern and Southern Africa.

Homo erectus emigrated out of Africa about 1.8 million years ago and was probably the first *Homo* species to so venture. Two species were derived from *H. erectus. H. antecessor* groups traveled to Europe, and *H. heidelbergensis* groups spread to East Africa and Eurasia. The latter two species gave rise to the Neanderthals and Denisovans through mutation changes rather than conquest and replacement. The Neanderthals spread through Europe and the Near East, while the Denisovans migrated through Central and East Asia to Southeast Asia and Oceania.

Homo sapiens appeared in Africa about 300,000 years ago and dispersed throughout the continent. Between 130,000 and 115,000 years ago, populations of *H. sapiens* (known as Cro-Magnon), migrated to the Near East and to

Europe, eventually reaching China, India, Southeast Asia, Oceania, New Guinea, Australia, Iran, Turkey, Spain, Portugal, Russia, Central Asia, Korea, Japan, Northern Europe, and eventually the New World. Along the way, they mated with the Denisovans, and the Neanderthals. This was disclosed by the detection of chromosomal fragments from the two ancient species in modern humans. Human travel into Alaska from Northeastern Asia occurred fifteen to twenty millennia ago. The Denisovans and Neanderthals eventually became extinct.

The history of the American continents begins with the Paleo-Indians. Migrants from Northeast Siberia crossed Beringia, a land area connecting Siberia to Alaska. Later, as the ice age ended and ocean waters rose, Beringia became the Bering Sea.

Some groups of Paleo-Indians settled in Alaska and northwest Canada. Others moved south along the Pacific Coast. Many traveled inland and settled in various locations in areas that would eventually become Canada in the colder north and the United States in the more southerly climate. They were hunter-gatherers, traveling in relatively small groups that formed numerous distinct tribes. The continent was vast, with a variety of climates, vegetation, animals, and soils that easily accommodated many such groups.

Eventually, in the more temperate areas, the tribes turned to agriculture, and formed many diverse and complex societies. In particular, the sub-tropical and tropical climates in Meso-America and South America encouraged the development of highly sophisticated civilizations such as the Inca in South America, the Olmec and Aztec in Mexico, and the Maya in Meso-America.

The agricultural development of the Americas was impressive, albeit with different species of plants and animals than in the Old World. Many of the crop plant species eventually were taken to and grown in the Old World. They included maize, tomatoes, potatoes, pumpkins, chili peppers, onions, strawberries, and vanilla.

Enter Christopher Columbus. The Americas were unknown late in the Fifteenth Century, except of course to their residents and perhaps to descendants of Leif Erikson and his followers, Norsemen from Iceland, who first discovered the North American continent in the Eleventh Century. Unlike the later visits by Columbus, the discovery did not inspire the Norse people to settle west of Greenland.

Constantinople fell to the Ottoman Empire in 1453, and the Silk Road to East Asia was closed to Christian traders. Interest began to develop for sailing westward across the Atlantic Ocean to reach Asian trading locations. Columbus was one of those interested. His proposal to King John of Portugal was rejected on the advice of the King's advisers that Columbus was wrong about the estimated distance, 2400 nautical miles, to the Far East. Also, sailing to the Far East by rounding the southern tip of Africa had already been done successfully.

Finally, Columbus won financial support from Queen Isabella and King Ferdinand of Spain. The monarchs presented Columbus with three ships: Niña, Pinta, and Santa Maria. He sailed in August 1492 and landed on an island of the Bahamas on October 12. Believing he had reached Asia, he called the inhabitants "indios". He made three more voyages, exploring the Lesser Antilles, Trinidad, and the North Coast of South America. Those trips and his subsequent years were not happy ones, for several reasons. His treatment of the natives was cruel, primarily due to their enslavement. He had a falling out with his royal sponsors. He also suffered with ill health and died at the age of 54.

Columbus's venture was followed by conquerors, priests, politicians, settlers, and writers from Spain, Portugal, France, Netherlands, and England in subsequent centuries. In 1519, Ferdinand Magellan of Portugal sailed around Cape Horn at the tip of South America and crossed the South Pacific Ocean. During a conflict with inhabitants of the Philippines, he was killed by a poisoned arrow.

The Europeans brought war, conquest, governance, and literature to the American continents. They also brought disease organisms to which they, the Europeans, were relatively resistant. Unfortunately, the natives were highly susceptible, and many of them died of diseases at considerably higher rates than from the invaders' weapons.

Except for a few Caribbean islands and small sections of the South American continent still under foreign control, all other European colonies became independent countries in subsequent years.

These few paragraphs sum up thousands of years of continuous growth and inventiveness, the tragedy of almost continuous warfare, the threat of global warming, and substantial population growth, but hopefully, also the desire and the means to save ourselves. We shall see.

History and geography are connected to each other in several ways. Geography is much more than printed maps. It is a group of sciences that study the various aspects of the lands: physical features, inhabitants, and characteristics of the Earth and, presumably, other planets and satellites in the Solar System, and perhaps eventually in other star systems. Geography's inclusion in this book is to examine its place as a contributor of threats to our survival discussed in the other essays.

What are the geographical sciences? There are two main branches: physical geography and human geography.

Each can be divided into several categories. Physical geography includes biogeography, climatology, coastal geography, environmental management, geodesy, geomorphology, glaciology, hydrology, landscape ecology, oceanography, paleogeography, soil science, and the present quaternary period science. Geodesy is a branch of science dealing with the shape and area of the earth and large portions of it.

Human geography includes cultural, developmental, economic, health, historical, political, religious, social, transportation, tourism, and urban geographies.

A critical effect of these geographies can be seen when human geographies and physical geographies are studied jointly, to show how their combined effects contribute to the threats to our survival. Consider, for example, an examination of coastal geography and climatology along with developmental geography and political geography and the combined threat of damage to our highly populated coastal lands and the drowning of low-lying island countries around the world. The sciences of climate change are now quite well known and demand immediate maximum mitigation. The current responses from many countries to the threat are not nearly enough to save the world from serious global warming. The final decisions are being made by political leaders who have been lobbied by energy companies that falsely claim the threat is over-rated and/or a hoax. The results have been inadequate responses to real threats, which will eventually destroy our precious, beautiful planet, unless we take immediate action to achieve maximum mitigation of the human activities that contribute to global warming. These problems are discussed more thoroughly in the essays on "Climate Change" and "`Environmental Pollution".

The state of glaciation is of vital importance to the

health and safety of the populations on earth. There have been five significant ice ages. Four occurred pre-historically: Huronian 2.4–2.1 billion years ago, Cryogenian 850–635 million years ago (mya), Andean-Saharan 460–430 mya, Karoo 360–260 mya, and Quaternary 2.6 mya–present. At present, glaciers exist in high mountain ranges and in the cold Arctic and Antarctic regions. If human induced global warming is not sufficiently controlled soon, most glaciers will melt, coastal cities will flood, and islands will drown as water from the thawed glaciers raises all the oceans.

At present the surface of the earth is home for eight billion people. The surface areas vary tremendously. Some areas are human-friendly habitats, while some others that are less than perfect can be modified in various ways, enabling people to live in them comfortably. Still other locations are not a bit friendly, and yet some people are forced to live there for various reasons. Poverty is a major culprit. Many poor people live on deserts, in jungles, on mountain tops, and other places that are very hot or very cold, or without adequate water supplies, or on poor land for growing crops.

There are other situations where human activity has been changing the environment, requiring drastic action to slow down or halt the changes. This is discussed earlier but bears repetition. The massive world forests are vital as carbon sinks to restrict the amount of carbon that enters the atmosphere. Unfortunately, many of the forests are being destroyed, especially in Brazil, Indonesia, and New Guinea, and will certainly hasten the heating up of the earthly atmosphere and surface. Atmospheric warming will have devastating effects.

One effect will be the thawing of glaciers in the polar areas and the mountains, as noted above. There is a second outcome that is not often discussed. The crop species that directly feed us and feed our animals, which provide us with meat and dairy products, are dependent upon the climates that they have been bred in for survival and to maintain productive abilities. Massive breeding programs will be required to develop new adaptive abilities for many crop and animal species.

I can bring personal experiences to provide an example of the harmful effect of warm temperatures on the development of a crop. For forty-six years I was a lettuce breeder for the U.S. Department of Agriculture in Salinas, California. In the United States lettuce is grown all year, primarily in California and Arizona. Smaller acreages are grown in Texas, Colorado, New Mexico, Florida, New York, Washington, and several other states. The types grown include iceberg, romaine, butterhead, and red and green leaf varieties. The Salinas Valley in California is the largest single lettuce production area in the nation. Iceberg lettuce is grown in the spring, summer, and fall at different locations in the Valley when the daytime temperatures are cool, ranging from the mid 60s to the high 70s, and rarely warmer. This temperature range demands lettuce varieties that grow well to develop a desirable sized plant. Iceberg varieties adapted for production in Salinas, if planted in a Midwestern or Eastern state during the warmer summer will grow larger than normal, with undesirable soft fluffy heads. Conversely, lettuces bred for the higher summer temperatures of the East or Midwest will be too small for commercial use if grown in Salinas. This problem will become worse as warming continues.

As global warming progresses, farm crops will react in various ways. Plant scientists will have to get to work to enable the various species to cope with the changes in crop growth progress. The weaker the mitigation efforts the greater difficulty for the plants to grow properly. The changes might require certain crops to be moved to new growing areas. Plant breeders and other plant scientists will have to develop new varieties at a faster rate. Farmers may also have to adapt by finding planting locations with different soil types and weather patterns.

It also may be necessary to grow some crops in buildings to have more control over "the weather." Some aspects of geography may well change drastically. A combination of drastic mitigation and adaptation to historical changes in the climate, and therefore the weather, will be the order of the future. Adequate progress will be dependent upon heretofore difficult political changes and striving for real changes in our handling of the crisis to come.

16

Transportation

Transportation is a phenomenon exceptionally useful to humans at all age levels, for babies in carriages, youngsters on bicycles and skateboards, and adults on all the other forms described in this and other essays. It serves both the working people and the vacationing people of the world. It plays a part in war, climate change, population size, science, technology, belief systems, and government. It is strongly involved in communication, business, agriculture, aspects of inequality, history, and geography. Transportation enables humans, animals, and goods to move from one place to another to carry out desired goals. In keeping with one of the major themes of this essay collection, transportation can be a means to do good things or to carry out evil designs. Transportation exists in multiple forms over the entire surface of the earth as well as in the atmosphere above and the waters below the surface.

Some people consider the possibility, far in the future, of traveling beyond the solar system, beyond our galaxy, to other galaxies, at the speed of light or even faster. Is this possible? No one truly knows. According to the theory of the Big Bang, this phenomenon has occurred once, 13.8 billion years ago, at the beginning of time, space, and the Universe. The initial expansion may have taken about 10^{-32} seconds, at the end of which the volume of the universe had increased by a factor of 10^{78}. Those numbers are almost unbelievable. Even the starships on Star Trek don't move that rapidly. The only way humans can travel the great distance to the nearest galaxy would be on a vessel designed to support multiple generations. Perhaps we should start planning now to build and furnish such vessels.

Back to a more pedestrian discussion. Transportation plays a part, major or minor, in each of the activities described in the chapters featured in this book.

At the present time, most types of transportation contribute to climate change as sources of greenhouse gases, particularly carbon dioxide (CO_2), methane (CH_4), and nitrous oxide (N_2O). These gases absorb radiation from the sun and trap heat in the atmosphere, thus raising global temperatures. War and climate change are the two most dangerous phenomena threatening our existence in the near or distant future.

The importance of transportation to human life and activity is easily understood from the number of ways we travel from one place to another, by land, water, and air. The number of ways to travel on land are more than double those of water and air travel together. Land travel started and continues with walking, also known as shanks' mare. Other means, from ancient times to the present, include chariots, wagons, automobiles, buses, trucks, bicycles, tricycles, skates, skateboards, skis, motorcycles, and railroads. We can also travel on the backs of animals: horses, oxen, donkeys, mules, ponies, elephants, alpacas, llamas, emus, and ostriches.

Water travel takes place on inland rivers, streams, and lakes and on the seas, bays, and oceans of the world. Inland, people swim, and travel in rowboats, canoes, kayaks, sailboats, and motorboats. On the larger bodies of water, the vehicles used include surface ships, submarines, surfboards, and small boats, depending upon the location and purpose.

Air travel includes flights on airplanes, gliders, helicopters, autogyros, balloons, and dirigibles. At present, space travel has been limited to human trips to the moon, and vehicular flights to Mars and other planets. Humans also travel to the International Space Station (ISS). The ISS is a joint project of the United States, Russia, Japan, European Union, and Canada. It can carry eleven astronauts and cosmonauts. On board, it features the performance of biomedical research designed to benefit humanity. Included in the project's goals are: 1. Improved medical scanning technology, 2. Creation of new drugs, 3. Creation of components of artificial blood for animals, 4. Providing the opportunity for students to cooperate in conducting research, 5. Use of NASA's ECOSTRESS payload to monitor heat in cities, 6. Creating artificial retinas in the microgravity of space, and 7. Monitoring heat safety on Earth, and eight other projects. In addition, there are opportunities for students on Earth to communicate with the travelers.

Pursuit of the conflicts of modern warfare requires a variety of transportation modes, particularly when the scope of the war is as large as recent ones, especially World War II: employing tanks, trucks, and passenger vehicles on land; ships on or under the open seas; and various types of flying machines, such as fighters, bombers, helicopters, and passenger airplanes. We are beginning to employ drones (self-propelled maneuverable space vehicles), and if we manage to get ourselves into another major war, there is little doubt that they will become important weapons of that war and successive ones if we survive.

Various transportation methods have been employed more and more in warfare. In future conflicts, that employment will be continued and probably augmented with new types of transport. Once again it will be an example of good vs bad effects.

One form of transportation that we may not consider as such starts with the same syllable and refers to the movement of thoughts from one mind to another. The word is transmission and in one sense it out-performs standard means of transportation. Transmission is essentially instantaneous. Distance is nearly irrelevant. The tools of the trade include telephones, telegraphs, computers, radios, and televisions. The travel route is through the air, or by wire. And the concepts used are ideas, thoughts, voices, and/or fingers. The basic aims employed may be good or bad, just as with other transportation types.

Transmission may carry messages, such as warnings of flooding or fires. Telephones may be used to make threats or for friendly conversations. Radios and televisions are useful for entertainment, receiving good or bad news from anywhere on Earth, and for advertising products. They can also function on an educational level, ranging from quiz shows to lectures on science, current events, history, and other subjects. On a different level, propaganda is information, usually biased or misleading, designed to promote or publicize a political cause, often of a malign nature. To be effective, propaganda requires transmission.

One form of transportation that also falls under the category of transmission is the unmanned aerial vehicle, better known as the drone. It is a powered aircraft with no human aboard, thus avoiding dull, dirty, and dangerous military missions for humans. It was invented in 1849 as a balloon and further developed through the 19th and early 20th Centuries. Control technologies became less expensive and dangerous in the 21st Century, and use of drones expanded into non-military usage, such as aerial photography, precision agriculture, product deliveries, drone racing, and criminal endeavors such as smuggling. Drones are built in several sizes: small, medium, large, larger, and largest, ranging from less than 20 pounds to over 1320 pounds. They also vary in operating altitude, speed, range, degree of autonomy and maximum altitude.

17

Future of the World

This final essay is the place to return to the second question I asked in the first essay, and essentially repeated in subsequent ones: What can be done to make our world better? Sadly, I find it hard to be optimistic about the future. The sword over our heads is the prospect that the use and abuse of power by aspiring political and corporate leaders will continue to get progressively worse, together with an inexorable movement in the direction of a global disaster from which there may be no return. What sort of disaster? We must concern ourselves with the following: A nuclear war, resulting in immediate mass murder and destruction in places where the devices explode, and delayed disasters where widely dispersed radioactive fallout reaches the ground, or where global ice melts and the added water inundates most of the world's shorelines. Or an ever-widening wealth gap, possibly leading to worldwide economic collapse. It is not unlikely that we might absorb all three in a perfect planetary storm. After all, we have been waging war for thousands of years and devising increasingly lethal weapons. The seeds of global warming were planted during the industrial revolution, if not earlier. And we have been toying with inducing economic collapse at least since the invention of the stock market in the Sixteenth Century.

In our journey through history, the possessors of power and money have led the way and have failed miserably to make our world a better place for all, primarily because that has rarely been their desires or their goals. Those whose demands have included the betterment of the world and all its inhabitants have been pushed aside or barely tolerated. Nevertheless, let's examine the question.

An odd factor is the apparent success, so far, of the MAD doctrine. MAD stands for Mutually Assured Destruction, a concept of sorts, which seems to have kept the possessors of nuclear weapons from using them, perhaps in fear that an attack by one nation on another would bring on a holocaust that would lead to destruction of all parties.

Let's look at some other ideas that may be little more than wishful thinking. In 1992, political theorist Francis Fukuyama claimed in his book "The End of History and the Last Man" that at the end of the cold war, we began a worldwide march toward democracy in the absence of war. A glance at today's news reports tells us that the journey is, at best, on hold. Fukuyama has recently downplayed that idea.

In another scenario, it seems that in each generation, some folks, tired and discouraged, believe that the young people of the time will eventually save us all. Unfortunately, the young grow older, many become more conservative, and the promise disappears, to be replaced by a new possibly wishful hope for the next young generation. Finally, while many people look to religion for personal solace, some also believe that faith leads to peace and happiness, not only for them, but, somehow, for all. However, religion has too often been an excuse for hatred and war.

Consider a more political approach. Theoretically, we wish to elect and support only the best people to lead us, but our attempts to do that are often crushed by those with the perpetual quest for power. This opportunity of choice doesn't even exist in countries with hereditary assumption of leadership by kings and queens. In countries without royalty, the opportunity often gets lost when a military or civil group, hungry for power, snatches leadership. Consider what we have in our world now: dictatorships, autocratic governments under a single party, absolute monarchies, and virtual anarchies, along with political democracies. Quests for raw power are difficult to prevent or stop, and we have often been unsuccessful in trying to deal with the problem. How many voters even ask themselves the question: Is this candidate simply looking for an opportunity to exert power? And how many voters admire a candidate's relentless ambition and aggressiveness? The name Donald Trump pops up here, with his efforts to overturn the 2020 election of Joe Biden, and his obvious desire to do away with the guardrails of democracy and become the first leader of an American dictatorship. If this seems preposterous, consider his self-worship, the charges against him, and the violent ways of many of his followers.

If the reader wishes to explore similar behaviors of other contemporary leaders, consider Victor Orban of Hungary, Jair Bolsonaro of Brazil, Vladimir Putin of Russia, Recep Tayyip Erdogan of Turkey, Hun Sen of Cambodia, Nicolas Maduro of Venezuela, Xi Jinping of China, and the list goes on and on. In some of these countries, pressures from within and outside the countries have forced their leaders to reform or withdraw. Bolsonaro was defeated in an election in 2022 but may run again.

Considering the corporate abuse of capitalism, in the United States and other countries, another possible solution is a worldwide conversion to democratic socialism. Socialism has been successful in several small countries, but badly abused in communist and fascist countries, where the quest for power has been violent and has over-ridden the virtues of the economic system. Socialism can be as chancy as capitalism.

There is another major change we might consider. After World War II, during the years of fear and horror engendered by the nuclear blasts that destroyed two Japanese cities, a few visionary people: political experts and plain citizens alike, embraced the idea of a world federal government as the only effective means of preventing further wars. (Author's note—I was a member of United World Federalists during that period.) At that time, such a courageous step seemed quite possible, but eventually the idea faded away because of a chorus of excuses and denials: It's not practical because no country will give up its armies and the right to make war. We would lose our sovereignty. We would be burdened with more layers of government and more taxes. And so on. But despite these objections, several governments around the world expressed sympathy with the idea. The main opponents were, of course, the Soviet Union and the United States, the two most powerful countries in the world at that time. Both believed they had the most to lose. Three quarters of a century later, is world government still a possibility? The litany of objections is still with us: more taxes, more regulations, loss of our sovereignty to foreigners. However, nations on five continents have formed seventeen regional organizations with varying degrees of consolidation, all of which, theoretically, could follow paths to a true worldwide federal governmental organization. The European

Union is the most advanced supranational politico-economic union now.

Is it possible that a form of world government now may be one controlled by the enormous commercial enterprises whose power is already on a world scale, and who might love to gain political control? The United Corporations of the World? Is this an inviting prospect? It may be to some monster corporations, for whom there is never enough power and money. For most folks, probably not. It is not very likely that the corporate world would embrace this idea if it was proposed, anyway, since making money by selling goods and services has been by far their major concern. They prefer lobbying for tax reductions and the like.

The benefits of a planet at peace under a democratically achieved world federal government should be considered as a pathway to permanent world peace. Within a world federation, each country would retain the right to govern itself. The power to make war on each other would be gone. The world government's ministry of defense would control all weapons of war. Nuclear weapons and other weapons of mass destruction would be destroyed. A worldwide watchdog agency would assure that no country would have the ability to build a stockpile of those weapons. Each country would have a national militia or police force bearing small-scale weapons sufficient to deal with relatively minor domestic problems. In the United States, the Second Amendment to the Constitution would have to be respected in its entirety, by including the first thirteen words regarding the possible need for state militias, as the *only* constitutional reason for bearing a weapon. Other reasons would not be protected by the constitution. They would require proof of specific need.

The possible benefits for the planet would be enormous. Our world has been at war almost constantly in one place or another throughout recorded history. The cost to the U.S. of the recent wars in Iraq, Syria, Pakistan, and Afghanistan, solely as military expenditure, has been estimated at $8 trillion. The total costs in military and civilian lives, of physical and mental maiming, of buildings destroyed, and of damage to the earth itself, are enormous and shouldn't even be estimated in mere monetary value. In a world at peace, many of our financial and human resources could be directed towards fixing the world, starting with repair of the damage inflicted by constant wars.

Even during periods when a nation is not actively participating in a war, it is likely to be in some stage of planning for one. Defense preparations are likely to require a significant portion of a nation's total budget and be, in a real sense, a waste of money, and for many countries, not affordable. The money could be better used in peacetime activities, and would be so used in a world federation.

At this writing, in the year 2024, two wars are in progress. Russia, under dictator-president Vladimir Putin, attacked Ukraine with no provocation in February 2014, and again in February 2022. Ukraine has been supported with weapons by Western countries, and at this writing has strongly resisted its larger neighbor. On October 7, 2023, the other war began with a surprise attack by the militant Palestinian group Hamas against Israel in the Gaza strip, the latest war of a series that began with an attack on the new state of Israel in 1948. There is no end to these conflicts in sight.

We could also end neglect of the planet, especially by preventing global warming from continuing to worsen. We could even address the modification of those areas in the world unfriendly to human habitation and endeavor to offer more people decent environments in which to live and work. We could eliminate human poverty and do a better job of controlling disease. As this is being written, we are just coming out of the throes of a world-wide pandemic, known as Covid 19. It is caused by severe acute respiratory syndrome coronavirus 2 (SARS-CoV-2). There have been almost 800 million cases and almost 7 million deaths as of March 2023. Under a world government, there should be a better controlled and efficient treatment program possible.

There is one worldwide practice that has been a blight for the human race for thousands of years and must cease: the terrible, hateful human activity of slavery. To eliminate slavery requires the efforts of the good people of Earth, which might be more easily accomplished in a world federation.

With this history in mind, I should point out that one basic human characteristic will not automatically change under a world federal government. The billions of human genomes will still exist, at the same level of variation that exists now, with behavioral traits ranging from good to bad, widespread differences in intelligence, as well as variation in physical traits. However, we could devote ourselves to investigating means to increase leadership opportunities for benevolent types of people. Free of war-making opportunities, it may be possible for us to effect significant improvements. In the early years of the transition, non-democracies would still exist, and their citizens might well wish to participate in a transition to improve their lives, and the lives of those who have not enjoyed the benefits of human existence.

It is worth reminding ourselves that a world organization already exists. The United Nations was formed at the end of World War II. This was a second attempt at forming a world organization to oversee world affairs. After World War I, the League of Nations came into being in 1920. Both organizations laudably accomplished a portion of their goals. Unfortunately, this did not include the primary goal of preventing war. The League terminated at the end of WWII in 1946, when the United Nations became active. The United Nations has not confronted a world conflict, but has had to deal with a thirty year religious conflict in Northern Ireland, the Cold War, the Korean 'police action', the prolonging of the Viet Nam War, the Gulf War, the Arab Spring uprising, and other wars involving the USA, the USSR, Iraq, Iran, Afghanistan, Syria, Lebanon, Israel, Palestine, Libya, and other states in Africa and the Eastern Mediterranean. The current wars in Ukraine and Israel have been added recently. This list of wars should be frightening, especially when considering the awful prospects in the future, with more advanced weapons likely to become available.

Both the League of Nations and the United Nations have included cultural, health, and social groups. For example, the UN Educational, Scientific, and Cultural Organization (UNESCO), and the UN International Children's Emergency Fund (UNICEF). These, and several others have been effective as aids to various countries in need of help, advice, and guidance.

A particularly difficult problem now, and one likely to create further difficulty in planning a transition to a world federation is The International Criminal Court. The United States, Russia, China, and several other countries around the world have refused to allow their citizens to go before the Court when charged with criminal acts and behavior affecting other countries.

The problem is that these groups and the UN itself are advisory and have minimum means of punitive action when required. This problem exists because the UN is not a real world government and has no lawful governmental power to enforce its advice with true legal or military action. A world federal government would have these powers. It is vital to the survival of the world and its eight billion people that the transit to world government be accomplished before we are overcome by Armageddon, the last battle between good and evil.

This is an appropriate place to return to our discussion of genes. Can we turn the tables and cause trouble for our bad alleles? From that statement a word once again pops into the mind: eugenics. The history of eugenics is not a happy one. Is there any reason to believe that we can do better in a peaceful world with modern methods and increased knowledge? Possibly, but not easily. Why not? First, if we wish to improve genotypes, that improvement must be for all, or nearly all of us, not just for a chosen elite, by whatever measure we assign that term. That means up to eight billion people and counting, a daunting number. Second, if the goal is elimination of "inferior people", by whatever measure or method, that is genocide and would be a despicable choice. The problem is a difficult one, because there *are* eight billion genotypes in the world, and genotypic changes occur with each generation, so that many newborn persons would require their own makeovers. The treatments would be constant. Even if we could identify the allelic status for every gene and every behavioral trait, the task might be overwhelming. And finally, who would determine what the most desirable allelic combinations should be? My short list of behavioral traits reflects my own bias for the good and bad allelic forms. Others may look at them differently. A modern eugenics program would appear to have very little virtue from any point of view.

The use of intelligence information regarding eugenic applications was a major factor in the recent past regarding treatment of people with low intelligence scores. One of the notorious persons in this area was a Nobel prize winning physicist, William Shockley. Shockley turned from his Nobel research field to become a self-anointed expert in human genetics, a subject in which he had no training nor experience, and in eugenics, a non-science that qualifies as a belief system. He proposed that people with IQs below 100 should voluntarily agree to be sterilized and get paid for it. He also believed that black people were genetically inferior, a deficit that he believed could not be remedied by betterment of their environments. His statements and writings show that he was a racist and genetically ignorant. (See genetics discussion in the first essay.)

But let's not give up on genetics. Rather, let's consider genes in the context of gender. About half the people in this world have genomes that are different from those of the other half. Specifically, one half has two X chromosomes and no Y chromosome. The other half has one X and one Y. We call the first group female and the second male. The Y chromosome has alleles found on no other chromosomal pair. How much effect will those differences have on the distribution and function of behavior alleles?

Recent research results, especially those published in the last couple of decades, provide evidence that there is a genetic basis for many if not all differences between males and females affecting psychological, personality, and behavioral traits. Additionally, the sex hormones testosterone and estrogen occur in both males and females. Testosterone is primarily a male hormone, but also occurs at a lower level in females. Estrogen, primarily a female hormone, also occurs in smaller amounts in males. The levels of both hormones vary in both sexes. This suggests that hormone level itself is a quantitative trait and varies according to the number of alleles controlling expression of the trait in the same way as for behavior.

In fact, hormone level was once considered the primary cause of the gender differences between men and women. Now, scientists recognize that there are also direct genetic effects on behavioral and psychological differences, such as aggressiveness and antisocial behavior. We must do much more work to determine the degree of variation in genetic and environmental influences on men and women. Perhaps then we will have some room in our hearts and minds for optimism.

Despite our incomplete understanding of the overall genetics of behavior, and of gender differences regarding those genetics, I shall walk gingerly out on this limb: When the time arrives that at least half of the leaders in politics, business, academia, and most other fields requiring leadership, are female, I believe that the expression of behavioral benevolence will increase and may be sufficient to alter our direction and move us away from disaster and towards peace, a healthy planet, a sensible economic system, and other signs of intelligent behavior. Time will show us whether this tentative prediction will come true or is complete nonsense.

At this point, it is long overdue for me to discuss the place of women in history, in the world, and in the United States, from as far back as we can to the present modern times. It is a reasonable assumption that in pre-historic times the roles of men and women may have been similar in the small hunter-gatherer groups of those times. Women likely participated in animal hunting, artistry, and tool making along with giving birth. This proposal agrees with the activities of modern women when having the opportunity, long overdue, to participate in society affairs along with males. As civilization progressed, the bigger, stronger, but not necessarily superior males, took over.

The rights of property ownership, voting, access to jobs, participation in team sports, and others taken for granted for centuries by nearly everybody who wasn't female were slowly granted to women over the years. The landmark event in the USA was the 19th Amendment to the Constitution, ratified in 1920, that granted women the right to vote.

This is an appropriate place to emphasize a need to expand the above pronouncements to include other neglected groups. I refer to people in groups that are not white in the western world, and those who are neglected in the southern and southeastern Asian worlds. They include native Americans, blacks, and browns in the Americas, Muslims and Mongols in China, aborigines in Australia.

The path of history is essentially linear. All the phenomena that contributed to that path over the past ten thousand years have marched inexorably forward from their simple beginnings to their present states, including the nature of war, governance, religion, science, the arts, population, climate, the state of knowledge, travel, agriculture, business, and on and on. All the changes are the products of our genes. That will continue and does not always bode well.

So, what can we hope for? I believe that we must make three major changes in the pathway. One, we must change the nature of our governance by creating a world federal government to bring about world peace and halt the destruction of our planet. Two, within the world structure, we must bring about widespread democracy and its benefits of equality. Three, the participation of women in all aspects of leadership must continue to rise. This last statement should apply to all religious, ethnic, racial, gender, and any other groups whose present existence and ability to participate in world and national affairs is based only on their group membership rather than on their desires and abilities to participate in human affairs.

There is one caveat that should be considered regarding greater participation of women in leadership. Their participation has increased from almost non-existent to minor over the years. It is possible that when female participation becomes large in scale, the virtues that they would bring to positions of leadership might become more like disturbing masculine traits and less like feminine ones than at present. This might attenuate the female virtues of less aggressive ambition and greater social moderation. If one looks at some recent female leaders: Margaret Thatcher (Great Britain), Golda Meir (Israel), Angela Merkel (Germany), Ellen Johnson Sirleaf (Liberia), and Indira Gandhi (India), among many others, it's difficult to draw conclusions. Their personalities and behavioral traits have differed substantially, as with any other human group. We also see the unsatisfactory behavior in a few of the women in the Republican Party now serving in the U.S. Congress. This caveat applies to others described in the previous paragraph.

Are all three of these incredibly difficult changes possible? The odds against them occurring are large. As I have earlier noted, the effects of human genes are moderated by the natural environment and by the environments we have created. The natural environment has always been with us, although some aspects have changed over time. All the environments that we have created: in politics, education, science, industry, and culture have come into existence over many centuries, and have developed under the control of our genes. They have been modified throughout time but have remained part of the totality of the environmental influence upon us. We react to all those environments under the guidance of our genes. In the end, therefore, it is still the genes that control us, directly and indirectly. That spells trouble and calls for action to save our world.

www.ingramcontent.com/pod-product-compliance
Lightning Source LLC
Chambersburg PA
CBHW071828210526
45479CB00001B/41
* 9 7 9 8 3 3 3 5 1 7 4 8 7 *